Printed at the Mathematical Centre, 49, 2e Boerhaavestraat, Amsterdam.

The Mathematical Centre, founded the 11-th of February 1946, is a non-profit institution aiming at the promotion of pure mathematics and its applications. It is sponsored by the Netherlands Government through the Netherlands Organization for the Advancement of Pure Research (Z.W.O), by the Municipality of Amsterdam, by the University of Amsterdam, by the Free University at Amsterdam, and by industries.

MATHEMATICAL CENTRE TRACTS 60

F. GÖBEL

QUEUEING MODELS INVOLVING BUFFERS

MATHEMATISCH CENTRUM AMSTERDAM 1976

AMS(MOS) subject classification scheme (1970): 60K25

ISBN 90 6196 108 4

CONTENTS

PREFACE

This tract is a corrected version of my thesis [GÖBEL 1974]. The main change is in Chapter 2, where the proofs of theorems 2.1 and 2.2 have been corrected and simplified following suggestions by Dr. W. Vervaat, to whom I express my thanks for his constructive criticism.

I am indebted to Prof.dr. J.Th. Runnenburg for his guidance and cooperation during the preparation of the thesis, his contributions to the results, and his patience throughout the whole process.

CHAPTER 1

INTRODUCTION

1.1 THE FRAMEWORK

In this book we consider some queueing models, most of which can be considered as specific cases of the following situation.

Customers arrive at a single service channel, called "filling-line", carrying loads of varying sizes and of M different types, to be indicated by $j = 1,2,\dots,M$. Each customer carries one type only; we use the term "j-customer" if his load is of type j.
The n-th customer ($n \geq 1$) arrives at time [1] $\underline{y}_0 + \underline{y}_1 + \dots + \underline{y}_{n-1}$ where the variables $\underline{y}_0, \underline{y}_1, \dots$ are identically distributed, non-negative, and independent, with $\mathcal{E}\underline{y}_n = \lambda^{-1} < \infty$. We will frequently assume that the $\underline{y}_n$ are exponentially distibuted. Sometimes it is convenient to assume a customer numbered 0, who arrives at time 0.

The probability that the n-th customer is a j-customer is p_j ($j = 1,\dots,M$), independently of the arrival times and the types of the other customers. Occasionally, we use the abbreviation λ_j for λp_j.
The service operation consists in transporting the loads through the filling-line into *buffers* of given capacities. A customer leaves the system when this transport has been completed. There is only one filling-line. The switch-over time from one filling-operation to the next is 0.
The buffers, which can be emptied, can contain loads of only one type at a time. This is perhaps the most essential feature of our models. However, after a buffer has been emptied, it can be used for another type. Only one buffer can be emptied at a time. Filling and emptying a buffer can be done simultaneously.

Figure 1.1 shows some of the features of the situation.

We usually make the following assumption on the filling- and emptying-times. The emptying-time $\underline{s}_j$ of a load of type j ($j = 1,\dots,M$) has a distibution function which may depend on j:

$$P\{\underline{s}_j \leq s\} = S_j(s),$$

---

[1] Random variables will be underlined.

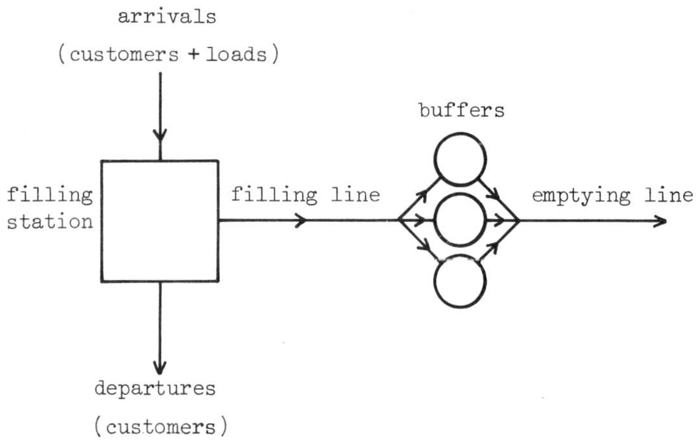

Figure 1.1

with $S_j(s) = 0$ for $s < 0$, and when the emptying-time is s, the filling-time is $\alpha_j s$ where α_j is a non-negative constant.
In some cases, we allow a more general relation between filling- and emptying-times.
When the sizes of the loads play a role, as is the case in finite buffers or in any question on the amount in the buffers, we usually assume that such sizes are proportional to the emptying-times.
The service discipline, i.e. the rule according to which we determine (a) the customer to be admitted to the filling-line, (b) the loads to be removed from the buffers, will be specified later. In many, but not in all, cases we will adopt the first-in-first-out discipline.

1.2 THE ORIGIN OF THE PROBLEM

Queueing occurs in a great variety of situations, and it always involves money, directly or indirectly. This commonplace highly applies to the field from which our problem originates. There, the customers are oil-tankers for which waiting is so expensive that very large investments to diminish their waiting-time are appropriate. For example, the problem how to choose the buffer sizes is an important problem.
The results presented in the following chapters hardly contribute to the solution of a practical problem of this kind but we hope that they are of interest apart from a possible application to the above-mentioned or a different area.

1.3 SUMMARY OF CHAPTERS 2 TO 5

The common feature of the models in chapter 2 is: the number of buffers, K, is equal to the number of types, M, and all buffers have infinite capacity.
In chapter 3, we consider one infinite buffer and M > 1.
In chapter 4, we consider K infinite buffers with 1 < K < M.
When the capacity of the buffers is infinite, we restrict our attention to "almost empty" buffers. This will be made precise later.
In the final chapter 5, some finite-buffer models will be discussed.
Sometimes we step outside the framework of §1.1, but in any case, our assumptions are stated at the beginning of each chapter. Most of the results are on the waiting-time.

1.4 SOME RELATED MODELS

The three most salient features of the class of models described in §1.1 are the following.

A) There are several types of customers.
B) Each customer requires two types of service, viz. filling and emptying.
C) The filling- and emptying-operation for one customer do not take place in series or parallel but are linked in a different manner.

A different although related class of models arises when instead of C), we impose the restriction that for each load, emptying can start only when the filling-operation has been completed. We then have two servers in series (with some complications), whereas the models of §1.1 are essentially single-server models.

The feature of *several types* is very essential in our models. Its consequences are exhibited most distinctly in §3.21, where the model is as simple as possible in other respects (one infinite buffer, service in order of arrival). It is true that we do allow finite filling-rates there, but a comparison of the results for finite and infinite filling-rates shows that this does not complicate the formulas to a great extent.

The relation between queueing-theory and the *theory of dams* is well-known. Our model is most reminiscent of dam theory when the number of types is 1, which is the case in §2.21, §5.3 and 5.4. Note that feature C) mentioned above arises in a dam-model as soon as the filling-rate is finite, which is quite a natural assumption to make. A model of this kind has been considered by Cohen (see [COHEN 1974]) who uses the term "gradual input".

A model with several types of customers in which the distribution of the service time of a customer depends on the type of his predecessor, has been considered by Gaver (see [GAVER 1963]). The dependence in question is due to the occurrence of an "orientation time" which the server needs whenever the type of the customer changes. As noted in [GÖBEL 1965], Gaver's model is simpler than ours since his orientation times do not accumulate.

1.5 SOME NOTATIONAL CONVENTIONS

As noted before, random variables will be denoted by underlined letters, usually latin lower-case, with or without subscripts. The corresponding capital is used for the distribution function (df, plural dfs), and the corresponding capital with a "cup" for the Laplace-Stieltjes transform (LST) of that df. For example, let $\underline{c}$ be a random variable. Then

$$C(x) = P\{\underline{c} \le x\},$$

$$\check{C}(\tau) = \mathcal{E} e^{-\tau \underline{c}}.$$

This convention overrules certain well-established conventions, but it has distinct advantages. We have not attempted to extend the convention to two-dimensional dfs. Instead of "the LST of the df of $\underline{c}$" we will usually speak of "the LST of $\underline{c}$".
The symbol $\triangleq$ means "has the same df as". The symbol □ denotes the end of a proof. The conditional probability of the event A, given B, is denoted by $P\{A|B\}$, and a similar convention applies to conditional expectation.
Finally, a sequence of symbols of the form $\{A(x) - [x:=y]\}$ has the same meaning as $\{A(x) - A(y)\}$. This convention is used to shorten $A(x) - A(y)$ when $A(x)$ is a complicated form (which will usually depend on numerous other quantities other than x, e.g. y). Some simple rules for the above notation are

$$\{\{A(x) - [x:=y]\} - [x:=z]\} = \{A(x) - [x:=z]\}$$

and

$$\{A(x) - [x:=y]\}B(y) = \{A(x)B(y) - [x:=y]\}.$$

In the second example, A and B may contain other variables than x and y, respectively, but B should not depend on x.

1.6 SOME SIMPLE RELATIONS

In the sequel, especially in Chs. 2 and 3, the following simple relations may be useful. Consider an infinite, empty buffer, and suppose a customer enters the system. If s is his emptying-time, then certain other quantities of interest can be expressed at once in s, α, ω, where ω is the emptying-rate (unit of amount per unit of time). The figure below gives this information in a convenient format.

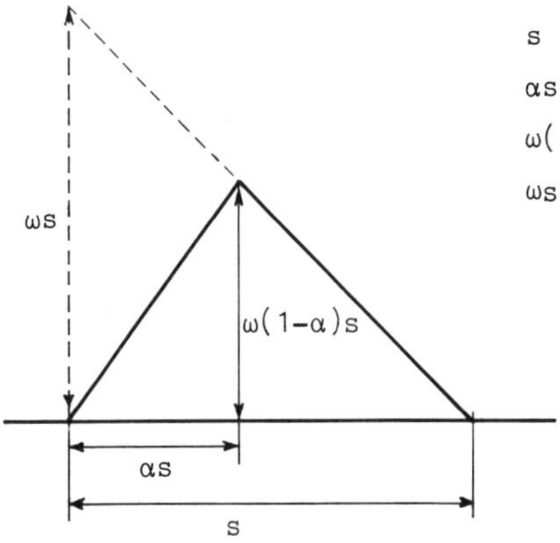

s	emptying-time
αs	filling-time
$\omega(1-\alpha)s$	height of maximum
ωs	size of load

Figure 1.2

Sometimes it is more convenient to express these quantities in terms of u, the filling-time.

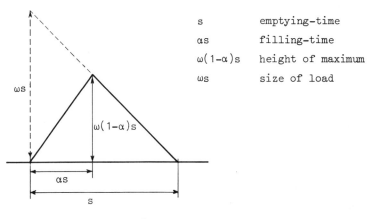

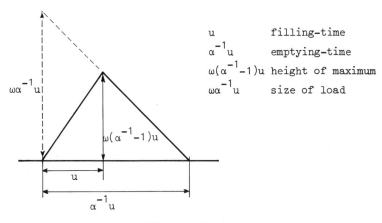

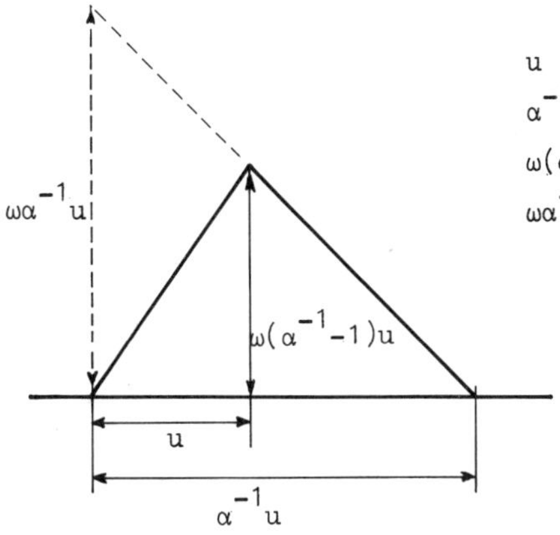

u	filling-time
$\alpha^{-1}u$	emptying-time
$\omega(\alpha^{-1}-1)u$	height of maximum
$\omega\alpha^{-1}u$	size of load

Figure 1.3

1.7 ORGANIZATION

We have used the decimal notation for the numbers of sections, so that e.g. §3.11 precedes §3.2.

Theorems, lemmas and figures are indicated by a chapter-number followed by a point, followed by a number which identifies the theorem etc. within the chapter. Formula-numbers follow the same convention and are enclosed in brackets.

CHAPTER 2

INFINITE BUFFERS; K = M

2.1 INTRODUCTION

When both the number of types of customers and the number of buffers are equal to M, it is possible to assign a buffer to each type. That is what we do in this chapter.

When, moreover, the buffers are infinite and almost empty, the behaviour of the customers will only be influenced by the filling operation, and not, for example, by the way in which the buffers are emptied. Hence, as far as the customers are concerned, we have a well-known queueing-model, viz. the G|G|1 model. In some cases, depending of course on the further assumptions one makes, more or less explicit results on the waiting-time, the queue length, and other quantities, can be found in the existing literature.

We will therefore consider not the customers but the loads, in particular the waiting-time of the loads, the amount in the buffers, and the so-called wet periods.

2.2 FILLING AND EMPTYING IN ORDER OF ARRIVAL

2.21 K = M = 1

If customers of one type fill one infinite buffer, explicit results on the amount in the buffer can be obtained.

The amount in the buffer immediately after the departure of the n-th customer will be denoted by $\underline{z}_n$; the amount at time t by $\underline{z}(t)$. As the strategy for emptying the buffer we choose: the emptying line is busy at time t when $\underline{z}(t) > 0$.

We consider, as usual, the process in the almost empty buffer, i.e. $\underline{z}(0)$ is finite with probability 1. We assume that $\underline{z}(0)$ has a given distribution.

The emptying-time for the n-th load will be denoted by $\underline{s}_n$, the size of the n-th load by $\omega\underline{s}_n$, and the filling-time by $\alpha\underline{s}_n$. We assume that

$$(2.1) \qquad \mu^{-1} = \mathcal{E}\underline{s}_n < \infty.$$

If $\alpha = 0$, the process $\underline{z}(t)$ is, in principle, the virtual waiting-time process for the G|G|1 queue, hence a well-known process, which we will not consider here.

If $\alpha \geq 1$, the process is rather trivial, although in a detailed treatment, several cases would have to be distinguished, none of which, however, is really interesting.

Hence we assume $0 < \alpha < 1$. A typical realization of the process is shown in figure 2.1.

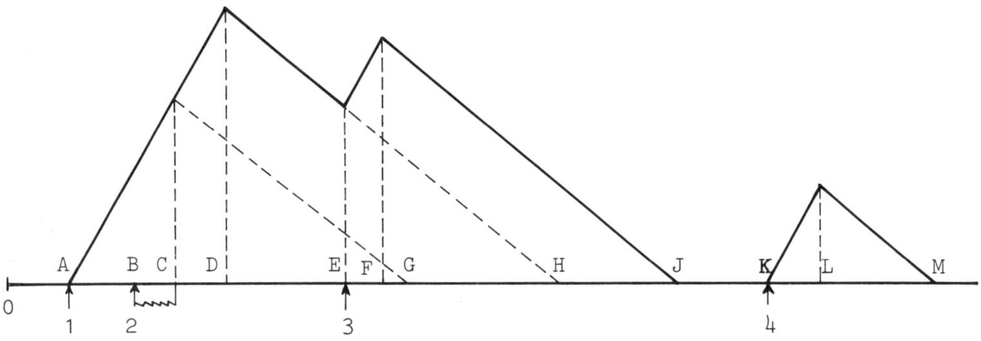

Figure 2.1

The arrows indicate the arrivals of customers 1, 2, 3, 4. Customer 1 stops filling at C and leaves the system. During the interval AG, the emptying-line is busy on 1's load. Customer 2 enters at B, has to wait a while, and starts filling at time C. At time D he leaves the system. Customer 3 does not wait, and at F he leaves the system. At time J the buffer becomes empty, etc.

The process $\underline{z}(t)$ contains an important imbedded process, viz. the relative maxima of the amount in the buffer. In many practical situations the maxima are of prime interest. In order to obtain information on these maxima, we consider the busy periods "induced by the filling operation". Following [COHEN 1974] we use the term "inflow periods". Such an inflow period starts when a customer who has not waited, starts to fill; it ends when a customer stops filling while no customer is waiting. In figure 2.1, the inflow periods are AD, EF, and KL.

At each moment when an inflow period ends, the amount in the buffer has a local strong maximum, and conversely. (By definition, the realization z(t) has a local strong maximum at $t = t_0$ if there exists an ε such that

$|t-t_0| < \varepsilon$ and $t \neq t_0$ imply $z(t) < z(t_0)$.)

Let $\underline{v}_n$ be the height of the n-th local strong maximum, $\underline{u}_n$ the length of the n-th inflow period, and $\underline{y}'_n$ the time from the end of the n-th inflow period until the first arrival of a customer. We assume that the interarrival times are *exponential*. Then it follows that $\underline{y}'_1, \underline{y}'_2, \ldots$ are also exponentially distributed, mutually independent, and independent of $\underline{u}_1, \underline{u}_2, \ldots$.

The time required to empty the buffer completely, starting with an amount $\underline{v}_n$, is equal to $\omega^{-1}\underline{v}_n$, provided no customer enters in the corresponding interval. Hence, if $\omega^{-1}\underline{v}_n \le \underline{y}'_n$, the next maximum $\underline{v}_{n+1}$ will be equal to $\omega(\alpha^{-1}-1)\underline{u}_{n+1}$. See also figure 2.2a.

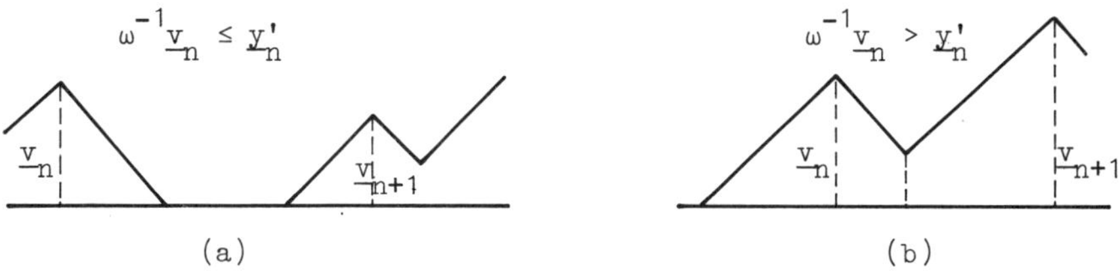

Figure 2.2

If $\omega^{-1}\underline{v}_n > \underline{y}'_n$, then there will be a positive amount $\underline{v}_n - \omega\underline{y}'_n$ left in the buffer at the arrival of the next customer (see figure 2.2b), and $\underline{v}_{n+1}$ is given by $\underline{v}_n - \omega\underline{y}'_n + \omega(\alpha^{-1}-1)\underline{u}_{n+1}$.

Summarizing these two cases, we have:

$$(2.2) \qquad \underline{v}_{n+1} = \omega(\alpha^{-1}-1)\underline{u}_{n+1} + \max(0, \underline{v}_n - \omega\underline{y}'_n).$$

If we write $\underline{v}_n = \omega\underline{w}_n + \omega(\alpha^{-1}-1)\underline{u}_n$, then (2.2) reduces to

$$(2.3) \qquad \underline{w}_{n+1} = \max(0, \underline{w}_n + (\alpha^{-1}-1)\underline{u}_n - \underline{y}'_n).$$

Note that in both cases, $\omega\underline{w}_n$ is the minimum amount in the buffer between the moments at which $\underline{v}_{n-1}$ and $\underline{v}_n$ are realized. Hence $\underline{w}_n$ is independent of the pair $(\underline{u}_n, \underline{y}'_n)$. Furthermore, the distribution of $\underline{u}_n$ does not depend on n (n ≥ 2). Hence, (2.3) has the Lindley structure, and we have at once the following result:

If $(\alpha^{-1}-1)\mathcal{E}\underline{u}_n < \mathcal{E}\underline{y}'_n$, $\underline{w}_n$ *has a limiting distribution with* LST

$$(2.4) \qquad \check{W}(\tau) = \frac{1-\rho^*}{1-\lambda\,\dfrac{1-\check{X}(\tau)}{\tau}} \qquad (\mathrm{Re}\,\tau > 0),$$

where $\check{X}(\tau)$ *is the* LST *of* $\underline{x}_n \overset{\mathrm{def}}{=} (\alpha^{-1}-1)\underline{u}_n$ $(n \geq 2)$, *and where* $\rho^* = \lambda\mathcal{E}\underline{x}_2$.

Since $\underline{u}_n$ is a busy period, $\check{X}(\tau)$ satisfies a Kendall-Takács equation (see c.g. [TAKÁCS 1962], p.58):

$$(2.5) \qquad \check{X}(\tau) = \check{S}((1-\alpha)\tau+\alpha\lambda-\alpha\lambda\check{X}(\tau)),$$

where $\check{S}$ is the LST of $\underline{s}_n$.

From (2.5) we find $(n \geq 2)$

$$\mathcal{E}\underline{x}_n = \frac{(1-\alpha)\mathcal{E}\underline{s}}{1-\lambda\alpha\mathcal{E}\underline{s}}$$

and

$$\mathcal{E}\underline{x}_n^2 = \frac{(1-\alpha)^2\mathcal{E}\underline{s}^2}{(1-\lambda\alpha\mathcal{E}\underline{s})^3},$$

and therefore, from (2.4):

$$\mathcal{E}\underline{w} = \frac{\lambda(1-\alpha)^2\mathcal{E}\underline{s}^2}{2(1-\lambda\alpha\mathcal{E}\underline{s})^2(1-\lambda\mathcal{E}\underline{s})}$$

where $\underline{w}$ is a random variable with LST $\check{W}(\tau)$.

Since $\underline{w}_n$ and $\underline{x}_n$ are independent, the limiting LST of $\underline{v}_n$ is given by

$$\check{V}(\tau) = \check{X}(\omega\tau)\,\check{W}(\omega\tau).$$

To conclude this section, we consider the *wet periods*. A wet period is a maximal open interval during which $\underline{z}(t) > 0$. Hence, when $0 \leq \alpha < 1$, a wet period may be considered as a busy period with respect to the emptying operation. It follows that the wet periods are independent of α as long as $0 \leq \alpha < 1$, and that they can be found by taking $\alpha = 0$. In fact, their LST $\check{B}$ is given by the functional equation

$$\check{B}(\tau) = \check{S}(\lambda+\tau-\lambda\check{B}(\tau)).$$

2.22 $K = M \geq 2$

Note that we allow general α's again.

We assume that the loads are taken out of the buffers in the order of filling (which coincides with the order of arrival of the corresponding customers). We assume that at time 0 a customer numbered 0, arrives.

Let $\underline{w}_n$ be the waiting-time of the n-th customer, i.e. the time from his arrival until the start of his filling operation. Let $\underline{x}_n$ be his filling-time, $\underline{s}_n$ the time required for removing the n-th load from the buffer if the emptying-line operates at full capacity, $\underline{y}_n$ the time between the n-th and (n+1)-st arrival times, $\underline{j}_n$ the type of the n-th customer, and $\underline{z}_n$ the waiting-time of the n-th load from the moment of arrival of the customer until the moment at which the removal of the n-th load from the buffer starts. For a pictorial summary of these definitions we refer to figure 2.3.

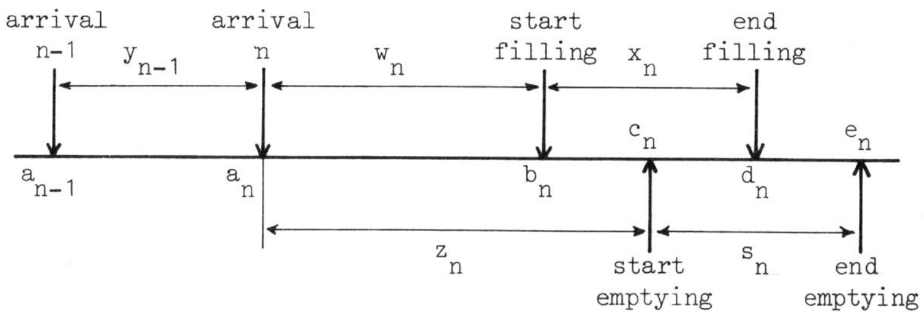

Figure 2.3

It is neither necessary nor desirable to take account of the types at this stage. Hence, the random variables $\underline{x}_n$ and $\underline{s}_n$ are mixtures of the corresponding conditional variables given the type. Note also that the original assumption that for each type, the filling-time is a constant times the emptying-time, is nowhere used in this section. Indeed, it is sufficient to assume that under the condition $\underline{j}_n = j$, $\underline{x}_n$ and $\underline{s}_n$ have a simultaneous distribution with all mass either in the first or in the second octant.

LEMMA 2.1.

(2.6) $$\underline{w}_{n+1} = \max(0, \underline{w}_n + \underline{x}_n - \underline{y}_n),$$

(2.7) $$\underline{z}_{n+1} = \max(0, \underline{w}_n + \underline{x}_n - \underline{y}_n,\ \underline{z}_n + \underline{s}_n - \underline{y}_n).$$

PROOF. For the n-th customer (or load), we define following epochs ($n \geq 1$).

$\underline{a}_n$: arrival time,
$\underline{b}_n$: start of filling,
$\underline{c}_n$: start of emptying,
$\underline{d}_n$: end of filling,
$\underline{e}_n$: end of emptying.

From the definition of $\underline{y}_n$ we have

$$\underline{a}_{n+1} = \underline{a}_n + \underline{y}_n.$$

The (n+1)-st customer starts to fill as soon as possible, that is: on his arrival, or else when the n-th customer has completed his filling operation. Hence

$$\underline{b}_{n+1} = \max(\underline{a}_{n+1}, \underline{d}_n).$$

Likewise we have

$$\underline{c}_{n+1} = \max(\underline{b}_{n+1}, \underline{e}_n).$$

Once a customer has started to fill the buffer, there are no restrictions on this operation, and we have simply

$$\underline{d}_n = \underline{b}_n + \underline{x}_n.$$

The emptying operation is completed at time $\underline{c}_{n+1} + \underline{s}_{n+1}$ *provided* the emptying-line has operated at full capacity. This condition is not necessarily fulfilled when the emptying occurs quicker than the filling for the (n+1)-st customer. As soon as the buffer becomes empty, the emptying will occur at the (slow) filling-rate, and the filling and emptying operations are completed simultaneously. Since there are no further restrictions on the emptying operation, we have

$$\underline{e}_n = \max(\underline{c}_n+\underline{s}_n,\underline{d}_n).$$

With the aid of the above five relations, we find

$$\underline{w}_{n+1} = \underline{b}_{n+1} - \underline{a}_{n+1} = \max(0,\underline{d}_n-\underline{a}_{n+1}) =$$

$$= \max(0,\underline{b}_n+\underline{x}_n-\underline{a}_n-\underline{y}_n) =$$

$$= \max(0,\underline{w}_n+\underline{x}_n-\underline{y}_n),$$

$$\underline{z}_{n+1} = \underline{c}_{n+1} - \underline{a}_{n+1} = \max(\underline{b}_{n+1}-\underline{a}_{n+1},\underline{e}_n-\underline{a}_{n+1}) =$$

$$= \max(\underline{w}_{n+1},\underline{e}_n-\underline{a}_n-\underline{y}_n) =$$

$$= \max(\underline{w}_{n+1},\underline{c}_n+\underline{s}_n-\underline{a}_n-\underline{y}_n,\underline{d}_n-\underline{a}_n-\underline{y}_n) =$$

$$= \max(0,\underline{z}_n+\underline{s}_n-\underline{y}_n,\underline{w}_n+\underline{x}_n-\underline{y}_n),$$

which completes the proof. □

When all filling-times are 0, the model is equivalent to the case $\alpha = 0$ of section 2.21. Hence, we may (and do) assume that for at least one type of customers, the value of α_j is not zero.

<u>LEMMA 2.2</u>. *If* $\underline{z}_0 \equiv 0$ *and* $\underline{w}_0 \equiv 0$*, then* $\underline{w}_n$ *has the same distribution as*

(2.8) $$\underline{w}'_n = \max_{0\le k\le n} \{\underline{v}_0+\ldots+\underline{v}_{k-1}\},$$

and $\underline{z}_n$ *has the same distribution as*

(2.9) $$\underline{z}'_n = \max_{0\le j\le k\le n} \{\underline{u}_0+\ldots+\underline{u}_{j-1}+\underline{v}_j+\ldots+\underline{v}_{k-1}\},$$

where $\underline{u}_i = \underline{s}_i-\underline{y}_i$, $\underline{v}_i = \underline{x}_i-\underline{y}_i$. *Furthermore:* $\underline{u}_0,\underline{u}_1,\ldots$ *are identically distributed;* $\underline{v}_0,\underline{v}_1,\ldots$ *are identically distributed; each set of* $\underline{u}$*'s and* $\underline{v}$*'s in which no index occurs more than once is a set of independent random variables.*

REMARK. The variables $\underline{u}_i$ introduced here have nothing to do with the inflow periods of §2.21.

PROOF. With the aid of (2.6) and (2.7) one can prove the following pair of relations by induction on n.

$$\underline{w}_n = \max_{0\le k\le n} \{\underline{v}_{n-k}+\ldots+\underline{v}_{n-1}\}, \tag{2.10}$$

$$\underline{z}_n = \max_{0\le j\le k\le n} \{\underline{v}_{n-k}+\ldots+\underline{v}_{n-j-1}+\underline{u}_{n-j}+\ldots+\underline{u}_{n-1}\}. \tag{2.11}$$

From the obvious fact that $\underline{u}_0,\underline{u}_1,\ldots$ are identically distributed and independent, as well as $\underline{v}_0,\underline{v}_1,\ldots$, it follows that one can renumber the variables in the right-hand sides of (2.10) and (2.11) to obtain (2.8) and (2.9). The last statement of the lemma follows at once from our assumptions on $\underline{y}_i$, $\underline{s}_i$ and $\underline{x}_i$. □

The following lemma goes back to a result of Runnenburg (see [RUNNENBURG 1960]); its present form is due to VERVAAT (personal communication, 1974). His proof is quite simple due to the fact that he considers the probability space Ω pointwise.

LEMMA 2.3. *Let* $\underline{u}_0,\underline{u}_1,\ldots$ *be identically distributed independent random variables, as well as* $\underline{v}_0,\underline{v}_1,\ldots$ with $\mathcal{E}\underline{u}_0 = \mu_1$, $\mathcal{E}\underline{v}_0 = \mu_2$, *and let*

$$\underline{M}_k = \max_{0\le j\le k} \frac{1}{k}(\underline{u}_0+\ldots+\underline{u}_{j-1}+\underline{v}_j+\ldots+\underline{v}_{k-1}).$$

Then

$$P\{\lim_{k\to\infty} \underline{M}_k = \max(\mu_1,\mu_2)\} = 1.$$

PROOF. Let

$$\underline{u}_0+\underline{u}_1+\ldots+\underline{u}_{k-1} = k\mu_1 + k\underline{\varepsilon}_1(k),$$
$$\underline{v}_0+\underline{v}_1+\ldots+\underline{v}_{k-1} = k\mu_2 + k\underline{\varepsilon}_2(k),$$

then $\underline{M}_k$ can be written as

$$\underline{M}_k = \max_{0\le j\le k} \left(\frac{j}{k}\mu_1+(1-\frac{j}{k})\mu_2+\frac{j}{k}\,\underline{\varepsilon}_1(k)+(1-\frac{j}{k})\underline{\varepsilon}_2(k)\right).$$

For almost all $\omega \in \Omega$ one has $\underline{\varepsilon}_1(k) \to 0$, $\underline{\varepsilon}_2(k) \to 0$ as $k \to \infty$, hence for each of these ω's

$$\frac{j}{k}\underline{\varepsilon}_1(k) + (1-\frac{j}{k})\underline{\varepsilon}_2(k)$$

converges to 0 uniformly in j as $k \to \infty$. Hence for these values of ω one has

$$\underline{M}_k = \underline{o}(1) + \max_{0 \le j \le k} (\frac{j}{k}\mu_1 + (1-\frac{j}{k})\mu_2) = \underline{o}(1) + \max(\mu_1,\mu_2) \quad (k \to \infty). \quad \square$$

<u>REMARK</u>. In Runnenburg's formulation of the lemma, the sequences $\underline{u}$ and $\underline{v}$ are required to be independent. It moreover gives only the $\limsup_{k\to\infty} \underline{M}_k$ instead of the limit.

The following two theorems deal with the limiting df

$$Z(z) \stackrel{\text{def}}{=} \lim_{n\to\infty} P\{\underline{z}_n \le z\}.$$

In [GÖBEL 1974], the proof of what is now Theorem 2.2 was rather clumsy, and strictly spoken, incorrect since it required the present, stronger, version of Lemma 2.3. Vervaat's proof means a considerable simplification, again made possible by pointwise considerations in Ω.

<u>THEOREM 2.1</u>. *If* $(\underline{z}_0,\underline{w}_0)$ *has an arbitrary df in the first quadrant, but is independent of the sequence* $\underline{u}_0,\underline{v}_0,\underline{u}_1,\underline{v}_1,\ldots$, *and if* $\mathcal{E}\underline{u} \ge 0$ *with* $u \not\equiv 0$ *or* $\mathcal{E}\underline{v} \ge 0$ *with* $\underline{v} \not\equiv 0$, *then* $Z(z) = 0$.

<u>PROOF</u>. First we consider the case $\underline{z}_0 = \underline{w}_0 = 0$.
Suppose $\mathcal{E}\underline{v} \ge 0$ and $\underline{v} \not\equiv 0$. From (2.6) and (2.7) we have $\underline{z}_n \ge \underline{w}_n$, hence

$$Z(z) = \lim_{n\to\infty} P\{\underline{z}_n \le z\} \le \lim_{n\to\infty} P\{\underline{w}_n \le z\}.$$

Now Lindley's analysis [LINDLEY 1952] shows that the latter limit does exist and is zero, hence $Z = 0$.

Suppose $\mathcal{E}\underline{u} \ge 0$ and $\underline{u} \not\equiv 0$. Since $\underline{x}_i \ge 0$, it follows from (2.9) that

$$\underline{z}_n \triangleq \underline{z}'_n \ge \max_{0\le j\le k\le n} \{\underline{u}_0+\ldots+\underline{u}_{j-1}-\underline{y}_j-\ldots-\underline{y}_{k-1}\}$$

$$= \max_{0\le k\le n} \{\underline{u}_0+\ldots+\underline{u}_{k-1}\} = \underline{t}_n \ ,$$

say, so that $Z(z) \le \lim\limits_{n\to\infty} P\{\underline{t}_n \le z\}$, and the same argument as above shows that $Z = 0$.

Next we consider the case $\underline{z}_0 = z_0$, $\underline{w}_0 = w_0$. It can be shown that $\underline{z}_{n+1}$ has the same df as $\underline{z}'_{n+1} = \max(\underline{A},\underline{B},\underline{C})$, where

$$(2.12)\qquad \begin{cases} \underline{A} = \max\limits_{0\le j\le k\le n-1} \{\underline{u}_0+\ldots+\underline{u}_{j-1}+\underline{v}_j+\ldots+\underline{v}_{k-1}\}, \\ \underline{B} = w_0 + \max\limits_{0\le j\le n} \{\underline{u}_0+\ldots+\underline{u}_{j-1}+\underline{v}_j+\ldots+\underline{v}_n\}, \\ \underline{C} = z_0 + \underline{u}_0 + \ldots + \underline{u}_n. \end{cases}$$

It follows that

$$P\{\underline{z}_n \le z \mid \underline{w}_0 = w_0, \underline{z}_0 = z_0\} \le P\{\underline{z}_n \le z \mid \underline{z}_0 = \underline{w}_0 = 0\},$$

hence, applying the result of the first case, we find that $Z = 0$.

Finally, in the case where $(\underline{z}_0,\underline{w}_0)$ has an arbitrary distribution in the first quadrant, it follows at once from Lebesgue's theorem on dominated convergence that again $Z = 0$. □

THEOREM 2.2. *If* $(\underline{z}_0,\underline{w}_0)$ *has an arbitrary df in the first quadrant, but is independent of the sequence* $\underline{u}_0,\underline{v}_0,\underline{u}_1,\underline{v}_1,\ldots,$ *and if* $\max(\mathcal{E}\underline{u},\mathcal{E}\underline{v}) < 0$, *then* $Z(z)$ *exists and is a df.*

PROOF. It can be shown that $\underline{z}_{n+1}$ has the same df as $\underline{z}'_{n+1} \stackrel{\text{def}}{=} \max(\underline{A}_n,\underline{B}_n,\underline{C}_n)$, where (cf. (2.12))

$$\underline{A}_n = \max_{0\le j\le k\le n-1} \{\underline{u}_0+\ldots+\underline{u}_{j-1}+\underline{v}_j+\ldots+\underline{v}_{k-1}\},$$

$$\underline{B}_n = \underline{w}_0 + \max_{0\le j\le n} \{\underline{u}_0+\ldots+\underline{u}_{j-1}+\underline{v}_j+\ldots+\underline{v}_{n-1}\},$$

$$\underline{C}_n = \underline{z}_0 + \underline{u}_0 + \ldots + \underline{u}_n.$$

Take a fixed ω for which $\lim\limits_{n\to\infty} u_n(w) = \mathcal{E}\underline{u}$ and $\lim\limits_{n\to\infty} v_n(w) = \mathcal{E}\underline{v}$. For such ω we have $C_n(\omega) \to -\infty$, $B_n(\omega) \to -\infty$ (on account of Lemma 2.3), and $\underline{A}_n \ge \underline{v}_0$ (take $j = 0$, $k = 1$ in the definition of $\underline{A}_n$), so that $\underline{z}'_{n+1} = \underline{A}_n$ for sufficiently large n. Since $\underline{A}_n$ is non-decreasing, it follows that

$$\underline{z} \stackrel{\text{def}}{=} \lim_{n\to\infty} \underline{A}_n = \lim_{n\to\infty} \underline{z}'_n$$

exists and is independent of the choice of $\underline{z}_0, \underline{w}_0$.

We now show that $\underline{z}$ is finite with probability 1.
For each n, let $\underline{k}_n$ be defined as the smallest integer with

$$\underline{A}_n = \max_{0\le j\le \underline{k}_n} \{\underline{u}_0+\ldots+\underline{u}_{j-1}+\underline{v}_j+\ldots+\underline{v}_{k_n-1}\}.$$

If $\{k_n(\omega)\}$ were unbounded, then, applying Lemma 2.3, $\{A_n(\omega)\}$ would contain a subsequence converging to $-\infty$, contradicting $\underline{A}_n \ge \underline{v}_0$. Hence $k_n(\omega)$ is bounded, hence $A_n(\omega)$ is bounded, hence $z(\omega)$ is finite.

Since we have strong convergence to $\underline{z}$, we have convergence in distribution. □

Now supposing that $\max(\mathcal{E}\underline{u}, \mathcal{E}\underline{v}) < 0$, the question arises whether Z can be determined. The general case (i.e. all α_j arbitrary) seems to be very hard. Only by considering the simultaneous df of $\underline{z}_n$ and $\underline{w}_n$, we obtained some result, namely a rather unpleasant integral equation for this df. We proceed as follows.

Using the notation of lemma 2.2, we write the equations (2.6) and (2.7) as follows:

$$\underline{w}_{n+1} = \max(0, \underline{w}_n+\underline{v}_n),$$

$$\underline{z}_{n+1} = \max(0, \underline{w}_n+\underline{v}_n, \underline{z}_n+\underline{u}_n).$$

If we define

$$F_n(z,w) = P\{\underline{z}_n \le z; \underline{w}_n \le w\}$$

for n = 1,2,..., then it can be shown that if z and w are ≥ 0, $F_n(z,w)$ satisfies

$$(2.13) \qquad F_{n+1}(z,w) = \int_{-\infty}^{\infty}\int_{-\infty}^{\infty} F_n(z-u, z\wedge w-v)\, d^2G(u,v),$$

where $z \wedge w$ denotes $\min(z,w)$, and where

$$(2.14) \qquad d^2G(u,v) = \sum_j p_j\ P\{u < \underline{u}_n \le u+du;\ v < \underline{v}_n \le v+dv \mid j_n = j\}.$$

The existence of

$$(2.15)\qquad F(z,w) = \lim_{n\to\infty} F_n(z,w)$$

can be shown as follows. If $\underline{z}_0 = \underline{w}_0 \equiv 0$, then the same renumbering-argument as in the proof of lemma 2.2 shows that the pair $(\underline{z}_n,\underline{w}_n)$ has the same distribution as the pair $(\underline{z}'_n,\underline{w}'_n)$ as defined in lemma 2.2. Hence, if $\underline{z}_0 = \underline{w}_0 \equiv 0$ then

$$F_{n+1}(z,w) = P\{\underline{z}_{n+1} \le z;\underline{w}_{n+1} \le w\}$$

$$= P\{\underline{z}'_{n+1} \le z;\underline{w}'_{n+1} \le w\}$$

$$\le P\{\underline{z}'_n \le z;\underline{w}'_{n+1} \le w\}$$

$$\le P\{\underline{z}'_n \le z;\underline{w}'_n \le w\} =$$

$$= P\{\underline{z}_n \le z;\underline{w}_n \le w\} = F_n(z,w).$$

Hence the limit $F(z,w)$ exists in that case.

It can be shown, in the same manner as in the previous theorem, that this limit exists also when the pair $(\underline{z}_0,\underline{w}_0)$ is given arbitrary deterministic start values or a df $F_0(z,w)$, provided the pair $(\underline{z}_0,\underline{w}_0)$ is independent of the sequences $\underline{u}_n$ and $\underline{v}_n$. It can also be shown that the limit is invariably equal to $F(z,w)$. By applying Lebesgue's theorem, it then follows from (2.13) that F satisfies the integral equation

$$(2.16)\qquad F(z,w) = \int_{-\infty}^{\infty}\int_{-\infty}^{\infty} F(z-u,z\wedge w-v)\,d^2G(u,v).$$

Furthermore, the function $F(z,w)$ given by (2.15) is a distribution function provided $\mathcal{E}\underline{u}$ and $\mathcal{E}\underline{v} < 0$, as a consequence of the following lemma.

<u>LEMMA 2.4</u>. *If* $F_n(x,y)$ *is a pointwise converging sequence of two-dimensional dfs, the marginal dfs of which converge pointwise to dfs, then* $F(x,y) = \lim_{n\to\infty} F_n(x,y)$ *is a df.*

<u>PROOF</u>. $F_n(x,y)$ is a df, hence $x < x'$ and $y < y'$ imply

$$F_n(x',y') - F_n(x,y') - F_n(x',y) + F_n(x,y) \geq 0.$$

If we let $x' \to \infty$ and $y' \to \infty$, we obtain

$$1 - F_n(x,\infty) - F_n(\infty,y) + F_n(x,y) \geq 0,$$

or, employing $G_n(x)$ and $H_n(y)$ as notations for the marginal dfs:

$$1 - F_n(x,y) \leq \{1-G_n(x)\} + \{1-H_n(y)\}.$$

Hence,

$$1 - F(x,y) = \{1-F_n(x,y)\} + \{F_n(x,y)-F(x,y)\} \leq$$

$$\leq \{1-G(x)\} + \{G(x)-G_n(x)\} + \{1-H(y)\} +$$

$$+ \{H(y)-H_n(y)\} + \{F_n(x,y)-F(x,y)\}.$$

Now $1 - G(x)$ and $1 - H(y)$ can each be made $< \varepsilon$ by choosing x and y sufficiently large. The remaining three terms can then each be made $< \varepsilon$ by choosing n sufficiently large. Hence $\lim_{x,y\to\infty} F(x,y) = 1$.

Continuity "from the right" can be shown as follows. Let $x \leq x'$ and $y \leq y'$. Then

$$F(x',y') - F(x,y) = F(x',y') - F(x',y) + F(x',y) - F(x,y)$$

$$\leq F(\infty,y') - F(\infty,y) + F(x',\infty) - F(x,\infty)$$

$$= H(y') - H(y) + G(x') - G(x),$$

and since G and H are continuous from the right, it follows that $F(x',y') - F(x,y)$ can be made arbitrarily small.

It is easily shown that the other conditions for F to be a df ($x < x'$ and $y < y'$ implies $F(x',y') - F(x,y') - F(x',y) + F(x,y) \geq 0$; $F(-\infty,y) = 0$; $F(x,-\infty) = 0$) are satisfied. This proves the lemma. □

$F(z,w)$ is the *only* solution of (2.16) among dfs. This can be shown as follows. Suppose $F^*(z,w)$ is any df satisfying (2.16). Now take this F^* as the df of $(\underline{z}_0,\underline{w}_0)$. Then, according to (2.13), F will be the df of all pairs $(\underline{z}_n,\underline{w}_n)$, and the sequence $F_0,F_1,F_2,\ldots$ converges trivially to F^*, hence F^* is equal to the (unique) limit F.

In summary, the limit $F(z,w)$ is a df which satisfies the integral equation (2.16) with $d^2G(u,v)$ given by (2.14), and it is the *only* df which satisfies (2.16). However, we have not been able to solve (2.16); its only possible use seems to lie in a numerical determination of $F(z,w)$.

Here we leave the general case until further notice.

Suppose all $\alpha_i < 1$, or, more generally, $P\{\underline{x}_n \le \underline{s}_n\} = 1$ (for all n). It follows from (2.7) that in this case

$$\underline{z}_{n+1} = \max(0,\underline{z}_n+\underline{s}_n-\underline{y}_n).$$

Hence, when $\mathcal{E}\underline{u} < 0$, $\mathcal{E}\underline{v} < 0$, the LST of Z is given by

$$\check{Z}(\tau) = \frac{\tau(1-\lambda\mathcal{E}\underline{s})}{\tau-\lambda+\lambda\check{S}(\tau)} .$$

Although in this special case, the simultaneous df $F(z,w)$ is not required to determine $Z(z)$, it would be interesting to find $F(z,w)$. RUNNENBURG (oral communication) formulated this problem, and he solved it in the following manner:

Let $\max(\mathcal{E}\underline{u},\mathcal{E}\underline{v}) < 0$, and let $\check{F}(\sigma,\tau)$ and $\check{B}(\sigma,\tau)$ be defined by

$$\check{F}(\sigma,\tau) = \mathcal{E}e^{-\sigma\underline{w}-\tau\underline{z}},$$

$$\check{B}(\sigma,\tau) = \mathcal{E}e^{-\sigma\underline{x}-\tau\underline{s}}.$$

By using the method of collective marks and analytic continuation he showed that

$$(2.17)\quad \begin{cases} \check{F}(\sigma,\tau)\{\lambda\check{B}(\sigma,\tau)-(\lambda-\tau-\sigma)\} = \\ = \tau\,\dfrac{\lambda-\tau-\sigma}{\lambda-\tau}\,\check{F}(0,\lambda)\check{B}(0,\lambda) + \lambda\,\dfrac{\sigma}{\lambda-\tau}\,\check{F}(\lambda-\tau,\tau)\check{B}(\lambda-\tau,\tau) \end{cases}$$

for all σ, τ with $\tau \ge 0$, $\sigma+\tau \ge 0$. The left-hand side of (2.17) is 0 when

$$(2.18) \qquad \lambda\check{B}(\sigma,\tau) = \lambda-\tau-\sigma$$

which may be considered as an equation in σ with τ as a parameter. The question whether (2.18) has a solution $\sigma = \xi(\tau)$ in a suitable region, can be answered affirmatively. (This depends on $\mathcal{E}\underline{x} \leq \mathcal{E}\underline{s}$, and this follows from $P\{\underline{x} \leq \underline{s}\} = 1$).

By substituting $\sigma = \xi(\tau)$ into the right-hand side of (2.17) he then obtained an expression for $\check{F}(\lambda-\tau,\tau)$, which is then again substituted into (2.17). The final result is

$$(2.19) \qquad \check{F}(\sigma,\tau) = \frac{\tau(1-\lambda\mathcal{E}\underline{s})\{\xi(\tau)-\sigma\}}{\xi(\tau)\{\lambda\check{B}(\sigma,\tau)-(\lambda-\tau-\sigma)\}} .$$

By series expansion at $\sigma = \tau = 0$ one can then obtain mixed moments like $\mathcal{E}\underline{w}\,\underline{z}$ as an explicit function of (mixed) moments of the pair $\underline{x}$, $\underline{s}$. In particular:

$$(1-\lambda\nu)\mathcal{E}\underline{w}\,\underline{z} = \frac{\lambda}{2}\,\mathcal{E}\underline{x}^2\underline{s} + \frac{\lambda^2\nu_2\mathcal{E}\underline{x}\,\underline{s}}{2(1-\lambda\nu)} + \frac{\lambda^2\mu_2\nu_2}{4(1-\lambda\nu)} - \frac{(1-\lambda\mu)\lambda\nu_3}{6(1-\lambda\nu)} - \frac{(1-\lambda\mu)\lambda^2\nu_2^2}{4(1-\lambda\nu)^2}$$

where $\mu = \mathcal{E}\underline{s}$, $\mu_2 = \mathcal{E}\underline{s}^2$, $\nu = \mathcal{E}\underline{x}$, $\nu_2 = \mathcal{E}\underline{x}^2$, $\nu_3 = \mathcal{E}\underline{x}^3$.

To conclude this section, we consider the wet periods and the wet j-periods, to be defined below.

The moment t belongs to a *wet period* if and only if, by definition, there is a non-empty buffer at time t. Clearly, the wet periods do not change when we take all α_j equal to 0, and the problem of finding the distribution of the wet periods is a classical one. In fact, the result given at the end of section 2.21 remains valid provided we interpret $\check{S}(\tau)$ as $\sum_j p_j\check{S}_j(\tau)$.

By definition, the moment t belongs to a *wet-j-period* if and only if the j-buffer is not empty at time t. So a wet j-period starts when a j-customer starts to fill the empty j-buffer, and it ends when the j-buffer becomes empty.

What can we say about the distribution of the wet j-periods? In the first place, it is independent of the filling-process and hence independent of $\alpha_1,\alpha_2,\ldots,\alpha_M$, and we may choose $\alpha_1 = \alpha_2 = \ldots = \alpha_M = 0$. Without loss of generality we may further assume that M = 2, for, if we consider the wet 1-periods, all types different from type 1 may be identified.

A typical realization of the process is shown in figure 2.4. At the epochs A, B, C, F, J, K, M a customer arrives, and fills the buffer of his type at an infinite rate. The numbers near the peaks indicate the types.

Note that the arrival of a 2-customer at B causes a delay of the emptying operation for the 1-customer who arrives at C, so that at F, the 1-buffer

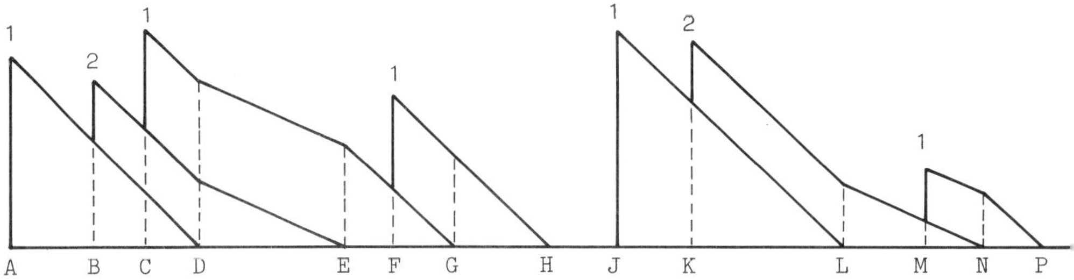

Figure 2.4

is not empty, and the wet 1-period which starts at A, ends at H. The next 1-period starts at J, and ends at L, in spite of the arrival of a 2-customer at K. The difference lies in the arrival at M, outside the service-time JL.

We consider an empty system, into which a 1-customer enters, who is served during the open interval $\underline{S}$ of length $\underline{s}$ (i.e. his emptying-time is $\underline{s}$). The customers, if any, who arrive in $\underline{S}$ are grouped to sequences which, by definition, are completed each time when a 1-customer enters. If no 1-customer arrives in $\underline{S}$, no sequence is formed. A 2-customer (named X) arriving during $\underline{S}$ belongs to a sequence if and only if there is a 1-customer who arrives later than X and before the end of $\underline{S}$. Hence, when we consider the types, the sequences are of the form 1 or 2,1 or 2,2,1 or 2,2,2,1 etc. Let $\underline{n}$ be the number of such sequences ($\underline{n} \geq 0$). Following a well-known argument (see [KENDALL 1951]), we now change the order of service as follows. When the initial 1-customer departs, we start to serve the customers of the first sequence as well as all customers who enter during the service-time of the customers of the first sequence, as well as all customers who enter during the service time of those customers, etc. Then we treat the second sequence and its "tail" in the same way, etc. Note that the completion of the wet 1-period is not influenced by this change of the order of service, since, by construction, there is always a 1-customer present when at least one sequence (or part of it) remains. The length $\underline{c}$ of the wet 1-period under consideration is thus given by

(2.20) $$\underline{c} = \underline{s} + \underline{c}'_1 + \underline{c}'_2 + \dots + \underline{c}'_{\underline{n}},$$

where $\underline{c}'_i$ ($i=1,2,\dots,\underline{n}$) denotes the duration of service on the i-th sequence and its "tail". Hence

(2.21) $$\underline{c}'_i \triangleq \underline{s}_1 + \underline{s}_2 + \dots + \underline{s}_{\underline{m}} + \underline{c},$$

where $\underline{s}_1, \dots, \underline{s}_{\underline{m}}$ are the service times of the 2-customers in the sequence, and where $\underline{m}$ is the number of 2-arrivals between two consecutive 1-arrivals.

Let μ_i be the expected emptying-time for the i-customers, and let $\rho = \lambda_1\mu_1 + \lambda_2\mu_2$. Then, if $\rho < 1$, one finds from (2.20) and (2.21):

$$\check{\Gamma}_1(\tau) = \check{S}_1(\tau+\lambda_1-\lambda_1\check{\Delta}_1(\tau))$$

and

$$\check{\Delta}_1(\tau) = \frac{\lambda_1\check{\Gamma}_1(\tau)}{\lambda-\lambda_2\check{S}_2(\tau)},$$

where $\check{\Gamma}_1$ and $\check{\Delta}_1$ are the LST of $\underline{c}$ and $\underline{c}'$, respectively.

As a check, one may verify that in the case of no 2-customers ($\lambda_2 = 0$, $\lambda = \lambda_1$), the equations become

(2.22) $$\check{\Gamma}_1(\tau) = \check{S}_1(\tau-\lambda_1-\lambda_1\Delta_1(\tau)),$$

(2.23) $$\check{\Delta}_1(\tau) = \check{\Gamma}_1(\tau),$$

which is equivalent to the Kendall-Takács functional equation:

(2.24) $$\check{\Gamma}(\tau) = \check{S}(\tau-\lambda-\lambda\check{\Gamma}(\tau)).$$

The same result is found if one admits 2-customers with service-times identically zero ($\check{S}_2(\tau) \equiv 1$).

For the expectation of $\underline{c}$ we find from (2.20) and (2.21)

$$(2.25) \qquad \mathcal{E}\underline{c} = \frac{\mu_1(1+\lambda_2\mu_2)}{1-\lambda_1\mu_1} .$$

Note that each epoch of any 1-period belongs to an ordinary wet period, hence, in a sense, the 1-periods are smaller than the wet periods. On the other hand, it is possible to choose λ_i and μ_i such that $\mathcal{E}\underline{c}$ is larger than $\mathcal{E}\underline{b}$. This somewhat paradoxical situation can be achieved quite easily, e.g. if $\lambda_1 = \frac{1}{9}$, $\mu_1 = 3$, $\lambda_2 = \frac{1}{3}$, $\mu_2 = 1$, then $\mathcal{E}\underline{c} = 6$, whereas (from 2.24)

$$\mathcal{E}\underline{b} = \frac{\mu}{1-\lambda\mu} = 4\tfrac{1}{2}.$$

2.3 EMPTYING THE BUFFERS WITH PRIORITIES

In Section 2.22, both the filling and the emptying operations take place in the order of arrival of the customers. We now consider a model in which the emptying order is different. It is not unusual that such a change makes the problem much more difficult. Therefore we simplify the model by assuming that the filling occurs infinitely fast. (The waiting-time of the customers is then identically 0). We now have, as it were, eliminated the buffers from the model, and an M|G|1-model remains with the special features of (a) different types of customers and (b) priorities.

The loads now play the role of the customers and "departure of a customer" has to be interpreted as "completion of the emptying operation on a load".

If we assign linear priorities to the M classes, we obtain a well-known problem; see e.g. [COBHAM 1954] and [KESTEN & RUNNENBURG 1957]. We do not consider these priorities here.

In many of the practical situations to which our model applies, it is quite natural to adopt a different form of priorities, which one might term *opportunist* priorities. This means the following:

(a) when a j-customer departs who leaves at least one j-customer waiting, the next customer to be served will be the j-customer who arrived first;

(b) when a j-customer departs who leaves no j-customer waiting, a discipline of sub-priorities will be used to determine which of the customers waiting, if any, will be served (unless M = 2, in which case no sub-priorities have to be defined);

(c) a customer who proceeds to the counter will be served immediately.

In the special case M = 2, these priorities have been termed *alternating*; they have been considered by various authors. In [AVI-ITZHAK, MAXWELL, and MILLER, 1965], a formula for the expected waiting-time is given, which coincides with a result in [GÖBEL 1969]. In [TAKÁCS 1968] and [GÖBEL 1969] the formulas for the generating functions f and g (see below) are derived, by different methods. Alternating priorities are also treated in Chapter 7 of [JAISWAL 1968].

In this section we consider, in the special case M = 2, the amount in the buffers and the wet periods for this model. We exclude service times that are identically zero.

The process is started as follows. To each set of integers k, a, b with k = 1 or 2, a ≥ 0, b ≥ 0, we assign non-negative numbers $p_{k,0}(a,b)$ with the following properties:

1) $\sum_k \sum_a \sum_b p_{k,0}(a,b) = 1$;

2) $a = 0$, $b > 0$ implies $p_{1,0}(a,b) = 0$;

3) $a > 0$, $b = 0$ implies $p_{2,0}(a,b) = 0$.

If $a+b > 0$, we define $p_{k,0}(a,b)$ as the probability that a 1-loads and b 2-loads are in the buffers at time 0, while priority is given to type k. If $a = b = 0$, we define $p_{k,0}(a,b)$ as the probability that no loads are present at time 0 and that the first customer to arrive is of type k.

Let $p_{j,n}(a,b)$ be the probability that the n-th *departing* customer is of type j and leaves a 1-loads and b 2-loads.

For complex x and y with $|x| \le 1$, $|y| \le 1$, we define the generating functions

$$f_n(x,y) = \sum_a \sum_b p_{1,n}(a,b)x^a y^b,$$

$$g_n(x,y) = \sum_a \sum_b p_{2,n}(a,b)x^a y^b.$$

The limits as $n \to \infty$ of these functions are given by the theorem below, but first we need some additional notation and a few lemmas.

Let $\mu_j = \mathcal{E}\underline{s}_j (j=1,2)$, $p = \lambda_1/\lambda$, $q = \lambda_2/\lambda$, $\rho_j = \lambda_j\mu_j (j=1,2)$, $\rho = \rho_1+\rho_2$.

LEMMA 2.5. (from [FELLER 1966], p.417). *If* $\check{B}_1$ *is the Laplace transform of a probability distribution with expectation* $0 < \mu_1 \leq \infty$, *and if* $\lambda_1 > 0$, *then the equation* $x = \check{B}_1(z+\lambda_1-\lambda_1 x)$ *has a unique solution* $x(z) \leq 1$, *and* x *is the Laplace transform of a distribution, which is proper if and only if* $\lambda_1\mu_1 \leq 1$, *defective otherwise.*

We apply this lemma with $z = \lambda_2-\lambda_2 y$ (keeping $0 \leq y \leq 1$), and we denote the solution of

$$(2.26) \qquad x = \check{S}_1(\lambda-\lambda_1 x-\lambda_2 y)$$

by $x = \xi(y)$. Similarly, we denote the solution of

$$(2.27) \qquad y = \check{S}_2(\lambda-\lambda_1 x-\lambda_2 y)$$

by $y = \eta(x)$. Note the similarity between these equations, the equation (2.18), and the Kendall-Takács functional equation (e.g. (2.24)).

Let $\gamma(x) = \xi(\eta(x))$, $\delta(y) = \eta(\xi(y))$, and let γ^n and δ^n denote iterates of γ and δ, respectively.

LEMMA 2.6. *If* $\rho < 1$, *then the sequence* $y, \delta(y), \delta^2(y), \ldots$ *is strictly increasing for all* $y \in [0,1)$.

PROOF. It is sufficient to show that $y < \delta(y) < 1$ for $y < 1$. We start by noting that

$$\check{S}_1(\tau) = \mathcal{E} e^{-\underline{s}_1\tau} \geq \mathcal{E}(1-\underline{s}_1\tau) = 1-\mu_1\tau,$$

so, from (2.26):

$$\xi(y) = \check{S}_1(\lambda-\lambda_1\xi(y)-\lambda_2 y) \geq 1-\mu_1(\lambda-\lambda_1\xi(y)-\lambda_2 y).$$

This yields the useful inequality

$$1 - \xi(y) \leq \frac{\lambda_2\mu_1}{1-\rho_1}(1-y).$$

Similarly

$$1 - \eta(x) \le \frac{\lambda_1\mu_2}{1-\rho_2}(1-x).$$

Taking $x = \xi(y)$ in the last inequality, we obtain

$$1 - \delta(y) \le \frac{\rho_1\rho_2}{(1-\rho_1)(1-\rho_2)}(1-y).$$

From $\rho < 1$ it follows that $\rho' = \dfrac{\rho_1\rho_2}{(1-\rho_1)(1-\rho_2)} < 1$, hence

$$1 - \delta(y) \le \rho'(1-y) < 1-y$$

(provided $y < 1$), hence $y < \delta(y)$.

Finally, $\delta(y) < 1$ follows from $\delta(y) = \eta(\xi(y))$, $\xi(y) > 0$, and the fact that each value of η is a value of $\check{S}_2$, which is the LST of $\underline{s}_2 \not\equiv 0$. □

<u>LEMMA 2.7</u>. *If* $\rho < 1$, *then* $\lim_{N\to\infty} \delta^N(y) = 1$ *for all* $y \in [0,1)$.

<u>PROOF</u>. Let $y \in [0,1)$. The sequence $y, \delta(y), \delta^2(y), \ldots$ is bounded from above by 1. Hence, the limit exists on account of the previous lemma, and it is ≤ 1. However, a limit < 1 is impossible on account of $y < \delta(y)$ in the interval $[0,1)$. Hence the limit equals 1. For $y = 1$, the assertion in the lemma is trivial.

<u>THEOREM 2.3</u>. *If* $\rho < 1$, *the limits*

$$f(x,y) = \lim_{n\to\infty} f_n(x,y)$$

and

$$g(x,y) = \lim_{n\to\infty} g_n(x,y)$$

exist and are given by

$$f(x,y) = \frac{(1-\rho)\check{S}_1(\lambda-\lambda_1x-\lambda_2y)}{x-\check{S}_1(\lambda-\lambda_1x-\lambda_2y)} \cdot \tag{2.28}$$

$$\cdot \left[\sum_{n=0}^{\infty} \left\{1-p\xi(\delta^n(y))-q\delta^{n+1}(y)\right\} - \sum_{n=0}^{\infty} \left\{1-p\gamma^n(x)-q\eta(\gamma^n(x))\right\}\right]$$

and

$$(2.29)\qquad g(x,y) = \frac{(1-\rho)\check{S}_2(\lambda-\lambda_1x-\lambda_2y)}{y-\check{S}_2(\lambda-\lambda_1x-\lambda_2y)}\cdot$$

$$\cdot\left[\sum_{n=0}^{\infty}\left\{1-q\eta(\gamma^n(x))-p\gamma^{n+1}(x)\right\}-\sum_{n=0}^{\infty}\left\{1-q\delta^n(y)-p\xi(\delta^n(y))\right\}\right].$$

OUTLINE OF PROOF. With the aid of the method of collective marks (or otherwise) one can derive the functional equations

$$(2.30)\qquad xf_{n+1}(x,y) = \check{S}_1(\lambda-\lambda_1x-\lambda_2y)\cdot$$

$$\cdot\left[f_n(x,y)-f_n(0,y)+g_n(x,0)-g_n(0,0)+px\{f_n(0,0)+g_n(0,0)\}\right],$$

$$(2.31)\qquad yg_{n+1}(x,y) = \check{S}_2(\lambda-\lambda_1x-\lambda_2y)\cdot$$

$$\cdot\left[g_n(x,y)-g_n(x,0)+f_n(0,y)-f_n(0,0)+qy\{f_n(0,0)+g_n(0,0)\}\right].$$

Letting $n \to \infty$, one can show in the same way as in section 4 of [KESTEN & RUNNENBURG 1957] that the limits f and g exist provided $\rho < 1$, and that f and g satisfy the limiting equations

$$(2.32)\qquad \{x-\check{S}_1(\lambda-\lambda_1x-\lambda_2y)\}f(x,y) = \check{S}_1(\lambda-\lambda_1x-\lambda_2y)\cdot$$

$$\cdot\left[-f(0,y)+g(x,0)-g(0,0)+px\{f(0,0)+g(0,0)\}\right],$$

$$(2.33)\qquad \{y-\check{S}_2(\lambda-\lambda_1x-\lambda_2y)\}g(x,y) = \check{S}_2(\lambda-\lambda_1x-\lambda_2y)\cdot$$

$$\cdot\left[-g(x,0)+f(0,y)-f(0,0)+qy\{f(0,0)+g(0,0)\}\right].$$

Since f(x,y) is analytic for $|x| < 1$, $|y| < 1$, and since the other factor in the left-hand side of (2.32) is 0 when $x = \xi(y)$, it follows that the right-hand side of (2.32) is 0 for $x = \xi(y)$:

$$(2.34)\qquad -f(0,y)+g(\xi(y),0)-g(0,0)+p\xi(y)\{f(0,0)+g(0,0)\} = 0$$

and similarly

$$(2.35) \qquad - g(x,0) + f(0,\eta(x)) - f(0,0) + q\eta(x)\{f(0,0)+g(0,0)\} = 0.$$

In particular we may substitute $x = \xi(y)$ in the last identity, to obtain:

$$(2.36) \qquad - g(\xi(y),0) + f(0,\delta(y)) - f(0,0) + q\delta(y)\{f(0,0)+g(0,0)\} = 0.$$

Adding (2.36) to (2.34), we obtain

$$- f(0,y) + f(0,\delta(y)) - \{f(0,0)+g(0,0)\}\{1-p\xi(y)-q\delta(y)\} = 0.$$

Since this relation holds not only for y, but for arbitrary iterates $\delta^n(y)$ of y, we obtain, by summation over n from 0 to N:

$$(2.37) \qquad f(0,y) = F(0,\delta^{N+1}(y)) - \{f(0,0)+g(0,0)\} \sum_{n=0}^{N} \{1-p\xi\delta^n(y)-q\delta^{n+1}(y)\}$$

$$(2.38) \qquad g(x,0) = g(\gamma^{N+1}(x),0) - \{f(0,0)+g(0,0)\} \sum_{n=0}^{N} \{1-p\gamma^{n+1}(x)-q\eta(\gamma^n(x))\}.$$

Now the series $\sum_0^\infty\{1-p\gamma^{n+1}(x)-q\eta(\gamma^n(x))\}$ is convergent provided $\rho < 1$, for the n-th term can be written as

$$p\{1-\gamma^{n+1}(x)\} + q\{1-\delta^n(\xi(x))\}.$$

For convergence it is sufficient to show that for all $x \in [0,1)$ both $1-\gamma(x)$ and $1-\delta(x)$ are $\le \rho'(1-x)$ for some $\rho' < 1$. As we have seen earlier, we may in fact choose

$$\rho' = \frac{\rho_1\rho_2}{(1-\rho_1)(1-\rho_2)} .$$

Hence we may let $N \to \infty$ in (2.37) and (2.38), and since f and g are continuous, we have

$$(2.39) \qquad f(0,y) = f(0,1) - \{f(0,0)+g(0,0)\} \sum_{n=0}^{\infty} \{1-p\xi(\delta^n(y))-q\delta^{n+1}(y)\},$$

$$(2.40) \qquad g(x,0) = g(1,0) - \{f(0,0)+g(0,0)\} \sum_{n=0}^{\infty} \{1-p\gamma^{n+1}(x)-q\eta(\gamma^n(x))\}.$$

Substituting (2.39) and (2.40) into (2.32), we find

$$(2.41)\qquad \{x-\check{S}_1(\lambda-\lambda_1x-\lambda_2y)\}f(x,y) = \check{S}_1(\lambda-\lambda_1x-\lambda_2y)\cdot$$

$$\cdot\Big[-f(0,1)+g(1,0)-g(0,0)+\{f(0,0)+g(0,0)\}\Big\{px +$$

$$+\sum_0^\infty\{1-p\xi(\delta^n(y))-q\delta^{n+1}(y)\} - \sum_0^\infty\{1-p\gamma^{n+1}(x)-q\eta(\gamma^n(x))\}\Big\}\Big].$$

Since the expression in the square brackets is 0 when $x = y = 1$, we have

$$(2.42)\qquad f(0,1) - pf(0,0) = g(1,0) - qg(0,0).$$

With the aid of this relation, (2.41) can be slightly simplified, and we obtain (2.28), with $f(0,0) + g(0,0)$ instead of $1-\rho$ as a factor in the right side. A proof of $f(0,0) + g(0,0) = 1-\rho$ can be given by comparing the present process with the process where the customers are served in order of arrival. A little reflection shows that if the n-th departing customer leaves no customers under one discipline, then the n-th customer leaves no customers under the other discipline. (The same is true for the arrival of a customer in an empty system.) Hence, since $f(0,0) + g(0,0) = 1-\rho$ under the "fifo" discipline, it follows that this relation holds in the present system *), and we obtain (2.28).

Formula (2.29) follows similarly, and the proof is complete. □

As an application consider $\underline{h}_n$, the amount in the 1-buffer at the departure of the n-th customer (i.e., at the completion of the emptying operation on the n-th load).

<u>THEOREM 2.4</u>. *If* $\rho < 1$, *then*

$$\check{H}_1(\tau) = \lim_{n\to\infty} \mathcal{E}e^{-\tau\underline{h}_n}$$

exists and is given by

$$\check{H}_1(\tau) = f(\check{S}_1(\omega_1\tau),1) + g(\check{S}_1(\omega_1\tau),1).$$

---

*) In [GÖBEL 1969] it was shown that $p\{f(0,0)+g(0,0)\} = (1-\rho)f(1,1)$; hence we now know that $p = f(1,1)$ and $q = g(1,1)$.

PROOF. Suppose the n-th customer leaves $\underline{a}_n$ 1-customers. Then

$$\mathcal{E}e^{-\tau\underline{h}_n} = \sum_a \mathcal{E}(e^{-\tau\underline{h}_n} \mid \underline{a}_n=a)P(\underline{a}_n=a) =$$

$$= \sum_a \mathcal{E}e^{-\tau(\underline{s}_1+\ldots+\underline{s}_a)\omega_1} \sum_b \{p_{1,n}(a,b)+p_{2,n}(a,b)\} =$$

$$= \sum_a \sum_b \left(\mathcal{E}e^{-\tau\omega_1\underline{s}_1}\right)^a \{p_{1,n}(a,b)+p_{2,n}(a,b)\} =$$

$$= f_n(\check{S}_1(\omega_1\tau),1) + g_n(\check{S}_1(\omega_1\tau),1).$$

If $\rho < 1$, the limit as $n \to \infty$ of the last member exists, hence the theorem. □

As a second application, consider the "queue length". Let $\underline{t}_n$ be the number of loads (irrespective of type) in the buffers at the moment when the emptying operation on the n-th load is completed.

THEOREM 2.5. *If* $\rho < 1$, *then*

$$D(x) = \lim_{n\to\infty} \sum_{t=0}^{\infty} P\{\underline{t}_n=t\}x^t$$

exists, and is given by $f(x,x) + g(x,x)$.

PROOF. Suppose $\underline{t}_n = \underline{a}_n+\underline{b}_n$, where $\underline{a}_n$ refers to 1-loads and $\underline{b}_n$ to 2-loads. Then a straightforward calculation shows that

$$\sum_0^{\infty} P\{\underline{t}_n=t\}x^t = f_n(x,x) + g_n(x,x),$$

and the theorem follows. □

Now we consider the process at a different sequence of epochs. When service on a 1-load starts and the load handled last is of type 2, then we say that a *1-period starts*. At such a moment, the number of 2-loads in the buffer is 0, and it may or may not have been 0 during an interval of positive length ending at that moment. The *start of a 2-period* is defined similarly.

Let $\underline{n}_j$ be the number of j-loads in the system just before a j-period starts in the stationary process.

Let

$$K_j(x) = \sum_{n=0}^{\infty} P\{\underline{n}_j=n\}x^n \qquad (j = 1,2)$$

THEOREM 2.6.

$$K_1(x) = \frac{g(x,0)-qg(0,0)}{g(1,0)-qg(0,0)},$$

$$K_2(y) = \frac{f(0,y)-pf(0,0)}{f(0,1)-pf(0,0)}.$$

PROOF. Define the events

$$B_1 = \{\text{a 1-period starts}\},$$
$$B_2 = \{\underline{n}_1=0\} \cap B_1,$$
$$B_3 = \{\underline{n}_1>0\} \cap B_1.$$

Then K_1 may be written as

$$K_1(x) = \sum_0^{\infty} P\{\underline{n}_1=n \mid B_1\}x^n.$$

Furthermore,

$$P\{\underline{n}_1=0 \mid B_1\} = \frac{P\{B_2\}}{P\{B_2\}+P\{B_3\}} = \frac{pg(0,0)}{pg(0,0)+g(1,0)-g(0,0)} =$$

$$= \frac{pg(0,0)}{g(1,0)-qg(0,0)}.$$

If n > 0, then

$$P\{\underline{n}_1=n \mid B_1\} = \frac{P\{\underline{n}_1=n \cap B_1\}}{P\{B_1\}} = \frac{p_2(n,0)}{g(1,0)-qg(0,0)},$$

and the formula for $K_1(x)$ follows. The formula for $K_2(y)$ follows by a suitable interchange of symbols. □

A closely related quantity is $\underline{n}_j^*$, the number of j-loads in the buffers just *after* a j-period starts. Let $K_j^*(x)$ be its generating function.

THEOREM 2.7.

$$K_1^*(x) = \frac{g(x,0)-(1-px)g(0,0)}{g(1,0)-qg(0,0)}.$$

PROOF. Obviously, $\underline{n}_1^* = \max(1,\underline{n}_1)$, and the theorem follows after a straightforward calculation. □

APPLICATION. Let $\underline{h}'$ be the amount in the 1-buffer just after a 1-period has started. Then a standard calculation shows that

$$\mathcal{E}e^{-\tau\underline{h}'} = K_1^*(\check{S}_1(\tau)).$$

APPLICATION. Let $\Phi_1(\tau)$ be the LST of the (maximal) intervals during which the 1-buffer is not empty. Then

$$\Phi_1(\tau) = K_1^*(\chi_1(\tau)),$$

where $\chi_1(\tau)$ is the LST of the wet-periods when the 2-customers are deleted from the process.

2.4 FILLING-PRIORITIES

When one adopts a filling-discipline other than 'first in, first out', one is faced with the difficulty of defining a reasonable emptying discipline. Let us briefly consider some possibilities. We start with a simple model.

A. Type 1 has preemptive filling priority over all other types; $\alpha_1 = 0$. Buffer 1 is emptied only when buffers 2,...,M are empty; the types 2,...,M are served in order of arrival.

In this case, we can apply the results of section 2.2 (or 2.1) to the types 2,...,M. The 1-customers have waiting-time 0. For the loads of type 1, we have an $M|G|1$ model with 'interrupted service', which is treated e.g. in [AVI-ITZHAK & NAOR 1963], model A.

B. Types 1,...,k ($1 < k < M$) have preemptive filling priority over the remaining types; $\alpha_1 = \ldots = \alpha_k = 0$. The loads of types 1,...,k are removed from their respective buffers in order of arrival when no loads of types k+1,...,M are present. The types k+1,...,M are served in order of arrival.

For the loads of types 1,...,k, the situation is more complicated than it is for the 1-loads in model A. However, it seems feasible to generalize the results of [AVI-ITZHAK & NAOR 1963], model A, in this direction.

C. Model A with 'non-preemptive' instead of 'preemptive'.

The waiting-time of the 1-customers may have a value > 0 now; its LST can be determined in a straighforward manner. We have not considered the 1-loads in this model.

D. Model A with '$0 < \alpha_1$' instead of '$\alpha_1 = 0$'.

Here the filling-process of types 2,...,M is interrupted by the 1-customers, and the emptying-process of the 1-customers is interrupted by the others. The model looks very complicated, but this seems inherent to each model in which filling-priority is given to a type with $\alpha \neq 0$.

CHAPTER 3

CUSTOMER INTERFERENCE IN ONE INFINITE BUFFER

3.1 INTRODUCTION

In this chapter, we assume one infinite buffer which is almost empty, and M types of customers (M > 1).

The n-th customer arrives at time $\underline{y}_0 + \ldots + \underline{y}_{n-1}$ $(n \geq 1)$ where each $\underline{y}_i$ has an exponential distribution with $\mathcal{E}\underline{y}_i = \lambda^{-1}$ and where $\underline{y}_0, \underline{y}_1, \ldots$ are mutually independent.

The type of the n-th customer is denoted by $\underline{j}_n$. We assume that p_j, defined by

$$p_j = P\{\underline{j}_n = j\} \qquad (j = 1,\ldots,M)$$

does not depend on n, that all $p_j > 0$, and that the variables $\underline{j}_1, \underline{j}_2, \ldots$ are mutually independent and independent of the arrival process.

Let the filling-time and the emptying-time for the n-th customer be denoted by $\underline{r}_n$ and $\underline{s}_n$, respectively. We assume that

$$P\{\underline{r}_n \leq \underline{s}_n\} = 1,$$

i.e. $F_j(r,s) = F_j(s,s)$ for $r \geq s > 0$. (For a discussion of this assumption see section 3.22). Further we assume that the pairs $(\underline{r}_1, \underline{s}_1), (\underline{r}_2, \underline{s}_2), \ldots$ are mutually independent, independent of the arrivals process, and that each $(\underline{r}_n, \underline{s}_n)$ is independent of $\underline{j}_1, \underline{j}_2, \ldots$ except (possibly) $\underline{j}_n$.

The waiting-time of the n-th customer is denoted by $\underline{w}_n$. We assume that $\underline{w}_1$ has a given distribution, which may depend on $\underline{y}_0$, but which is independent of the rest of the process.

The only emptying-strategy we consider in this chapter is: the emptying-line is busy whenever the buffer contains a positive amount.

3.2 FILLING IN ORDER OF ARRIVAL

In this section and its sub-sections, we assume that filling takes place in order of arrival and that the filling station is never idle when there is a customer waiting, unless the buffer contains goods of a type differing from the type of the first customer in the queue.

3.21 The waiting-time

In this subsection, we will obtain the transient behaviour of the waiting-time in principle, the stationarity condition, and the limiting behaviour in a more explicit form.

The method is roughly as follows. We consider, beside the process defined above ("the original process"), a modified process in which each customer who finds the emptying-line occupied, waits with filling until the emptying-line is idle, regardless of the type in the buffer. The modified process is fairly simple and serves as a basis for obtaining results for the original process.

First we state some lemmas.

LEMMA 3.1. [TAKÁCS 1962]. *In an* M|G|1 *queueing model, let* λ *be the density of the arrivals process,* $\check{S}(\tau)$ *the* LST *of the service times, and* $\check{W}_k(\tau)$ *the* LST *of the waiting-time of the* k-*th customer* ($k \geq 1$). *Then the generating function of* $\check{W}_k(\tau)$ *is given by*

$$\sum_{k=1}^{\infty} \check{W}_k(\tau)x^k = \frac{\tau x}{\lambda-\tau-\lambda x\check{S}(\tau)} \left\{\frac{(\lambda-\tau)\check{W}_1(\tau)}{\tau} - [\tau := \lambda-\lambda z]\right\} \tag{3.1}$$

where $|x| < 1$ *and* z *is the root with smallest absolute value of*

$$z = x\check{S}(\lambda-\lambda z). \tag{3.2}$$

We also need a slightly more general result, pertaining to the conditional waiting-time of the k-th customer, given that the preceding customer has a service-time with LST $\check{S}_j(\tau)$.

Denoting the LST of such a waiting-time by $\check{W}_k(\tau|j\cdot)$, we have the following lemma.

LEMMA 3.2. *In the above notation, the generating function of* $\check{W}_k(\tau|j\cdot)$ *is given by*

$$\sum_{k=1}^{\infty} \check{W}_{k+1}(\tau|j\cdot)x^{k+1} = \frac{\lambda\tau x}{\lambda-\tau} \left\{\frac{\check{S}_j(\tau)}{\tau} \sum_{1}^{\infty} \check{W}_k(\tau)x^k - [\tau := \lambda]\right\}.$$

PROOF. Using the method of collective marks, it is easy to establish

$$(3.3)\qquad \lambda\check{W}_k(\tau)\check{S}_j(\tau) = \tau\check{W}_k(\lambda)\check{S}_j(\lambda) + (\lambda-\tau)\check{W}_{k+1}(\tau|j\cdot).$$

(See also [RUNNENBURG 1965], p.402; our τ is RUNNENBURG's $\lambda-\lambda X$.) Multiplication by x^{k+1} and summation over k immediately gives the desired result.□

Although lemma 3.2 has a wider applicability, we apply it mainly to our model with $\check{S}_j(\tau)$ equal to the LST of the emptying-times of the j-customers. Hence

$$\check{S}_j(\tau) = \int_0^\infty e^{-\tau s} dF_j(\infty,s).$$

In the next lemma, we assume the conditions of our model, in particular on the way in which the types are assigned. We will denote by $\underline{s}$ an "unconditional" emptying-time, hence a random variable with df $\sum_j p_j F_j(\infty,s)$.

LEMMA 3.3. *If* $\lambda\mathcal{E}\underline{s} < 1$ *then the limits*

$$\check{W}(\tau|j\cdot) = \lim_{k\to\infty} \check{W}_k(\tau|j\cdot),$$

$$\check{W}(\tau) = \lim_{k\to\infty} \check{W}_k(\tau)$$

exist, and

$$\check{W}(\tau|j\cdot) = \frac{\lambda\tau}{\lambda-\tau}\left\{\frac{\check{S}_j(\tau)\check{W}(\tau)}{\tau} - [\tau := \lambda]\right\}.$$

PROOF. The statement about $\check{W}(\tau)$ is a well-known result. From formula (3.3) we obtain

$$\check{W}_{k+1}(\tau|j\cdot) = \frac{\lambda\check{W}_k(\tau)\check{S}_j(\tau)-\tau\check{W}_k(\lambda)\check{S}_j(\lambda)}{\lambda-\tau}$$

and by letting $k \to \infty$, we obtain both the existence of and the formula for $W(\tau|j\cdot)$. □

REMARK. If $\check{S}_j$ is independent of j in lemma 3.3, then $\check{W}(\tau|j\cdot) = \check{W}(\tau)$ for all τ and all j, and we may solve for $\check{W}(\tau)$, obtaining the Pollaczek-Khintchine formula, as is proper.

We return to the "original process". Consider the n-th customer ($n \geq 1$). Let $\underline{k}_n$ be defined as follows.

$$\underline{k}_n = \begin{cases} n \text{ if } \underline{j}_1 = \underline{j}_2 = \dots = \underline{j}_n, \\ \min \{k | k \geq 1; \underline{j}_{n-k} \neq \underline{j}_n\} \text{ otherwise.} \end{cases}$$

Let $\check{C}_n(\tau|j)$ be the conditional LST of the waiting-time of the n-th customer (in the original process) given that $\underline{j}_n = j$, and let $\check{C}_n(\tau|j;k)$ be the conditional LST of the same variable under the condition $\underline{j}_n = j$; $\underline{k}_n = k$. (In other words: the n-th customer is the k-th of a sequence of j-customers.) Then, of course,

$$(3.4) \qquad \check{C}_n(\tau|j) = \sum_{k=1}^{n} \check{C}_n(\tau|j;k)P\{\underline{k}_n=k|\underline{j}_n=j\} =$$

$$= \sum_{k=1}^{n-1} \check{C}_n(\tau|j;k)p_j^{k-1}(1-p_j) + \check{C}_n(\tau|j;n)p_j^{n-1}.$$

Let u be a complex number with $|u| < 1$. Then from (3.4) we have

$$\sum_{n=1}^{\infty} \check{C}_n(\tau|j)u^n = \Sigma_1 + \Sigma_2,$$

where

$$\Sigma_1 = \sum_{n=1}^{\infty} \check{C}_n(\tau|j;n)p_j^{n-1}u^n,$$

$$\Sigma_2 = \sum_{n=2}^{\infty} \sum_{k=1}^{n-1} \check{C}_n(\tau|j;k)p_j^{k-1}(1-p_j)u^n.$$

The sum Σ_1 can be reduced with the aid of lemma 3.1. Since the condition $\underline{j}_n = j$; $\underline{k}_n = n$ implies that all customers 1,2,...,n are of type j, the service-times of lemma 3.1 have to be interpreted as filling-times here, with df $F_j(r,\infty)$ and LST $\check{R}_j(\tau)$. We obtain

$$\Sigma_1 = \frac{\tau u}{\lambda-\tau-\lambda p_j u\check{R}_j(\tau)} \left\{\frac{\lambda-\tau}{\lambda} \check{C}_1(\tau|j) - [\tau := \lambda-\lambda z_j]\right\}$$

where z_j is the root with smallest absolute value of

$$z_j = p_j u\check{R}_j(\lambda-\lambda z_j).$$

Next we reduce Σ_2. The symbol q_j will be used as an abbreviation for $1-p_j$.

$$\Sigma_2 = \sum_{k=1}^{\infty} \sum_{n=k+1}^{\infty} \check{C}_n(\tau|j;k)p_j^{k-1}q_j u^n =$$

$$= \sum_{k=1}^{\infty} \sum_{m=1}^{\infty} \check{C}_{m+k}(\tau|j;k)p_j^{k-1}u^k q_j u^m =$$

$$= \sum_{m=1}^{\infty} q_j p_j^{-1} u^m \sum_{k=1}^{\infty} \check{C}_{m+k}(\tau|j;k)(p_j u)^k.$$

Again we apply lemma 3.1:

$$\Sigma_2 = \frac{\tau u q_j}{\lambda-\tau-\lambda p_j u\check{R}_j(\tau)} \sum_{m=1}^{\infty} u^m \left\{\frac{\lambda-\tau}{\tau} \check{C}_{m+1}(\tau|j;1) - [\tau := \lambda-\lambda z_j]\right\}$$

where z_j has the same meaning as in Σ_1.

Now we turn our attention to the modified process. Consider $\check{C}_{m+1}(\tau|j;1)$. By construction, the (m+1)-st customer is the first of a sequence of j-customers. Hence

$$\check{C}_{m+1}(\tau|j;1) = \check{H}_{m+1}(\tau|\underline{j}_m\neq j),$$

where $\check{H}$ refers to the modified process. Hence, using the abbreviation $\Delta = \lambda-\tau-\lambda p_j u\check{R}_j(\tau)$,

$$\Sigma_2 = \frac{\tau q_j}{\Delta}\left\{\frac{\lambda-\tau}{\tau} \sum_{m=1}^{\infty} \check{H}_{m+1}(\tau|\underline{j}_m\neq j)u^{m+1} - [\tau := \lambda-\lambda z_j]\right\}.$$

Since, from an elementary calculation,

$$\check{H}_{m+1}(\tau|\underline{j}_m\neq j) = \frac{\check{H}_{m+1}(\tau)-p_j\check{H}_{m+1}(\tau|j\cdot)}{q_j},$$

we obtain after some rearrangement

$$\Sigma_2 = \frac{\tau}{\Delta}\left\{\frac{\lambda-\tau}{\tau}\sum_1^{\infty} \check{H}_{m+1}(\tau)u^{m+1} - [\tau := \lambda-\lambda z_j]\right\} +$$

$$- \frac{p_j\tau}{\Delta}\left\{\frac{\lambda-\tau}{\tau}\sum_1^{\infty} \check{H}_{m+1}(\tau|j\cdot)u^{m+1} - [\tau := \lambda-\lambda z_j]\right\}.$$

If we take m = 0 in the first line of this formula, we have just Σ_1 (since $\check{H}_1(\tau) = \check{C}_1(\tau|j)$),

hence

$$\sum_{n=1}^{\infty} \check{C}_n(\tau|j)u^n = \Sigma_1 + \Sigma_2 = \Sigma_3 - \Sigma_4,$$

where

$$\Sigma_3 = \frac{\tau}{\Delta}\left\{\frac{\lambda-\tau}{\tau}\sum_1^{\infty} \check{H}_m(\tau)u^m - [\tau := \lambda-\lambda z_j]\right\},$$

$$\Sigma_4 = \frac{p_j\tau}{\Delta}\left\{\frac{\lambda-\tau}{\tau}\sum_1^{\infty} \check{H}_{m+1}(\tau|j\cdot)u^{m+1} - [\tau := \lambda-\lambda z_j]\right\}.$$

Let the LST of the emptying-times be denoted by $\check{S}(\tau)$. Hence

$$\check{S}(\tau) = \sum_j p_j\check{S}_j(\tau) = \sum p_j \int_0^{\infty} e^{-\tau s} dF_j(\infty,s).$$

Applying lemma 3.1 again, now to the modified process, we can rewrite the sum $\sum_1^{\infty}\check{H}_m(\tau)u^m$, occurring in Σ_3, as follows

$$\frac{\lambda-\tau}{\tau}\sum_1^{\infty} \check{H}_m(\tau)u^m =$$

$$= \frac{u(\lambda-\tau)}{\lambda-\tau-\lambda u\check{S}(\tau)}\left\{\frac{\lambda-\tau}{\tau}\check{H}_1(\tau) - [\tau := \lambda-\lambda z']\right\}$$

where z' is the root with smallest absolute value of

$$z' = u\check{S}(\lambda-\lambda z').$$

The sum $\sum_1^{\infty} \check{H}_{m+1}(\tau|j\cdot)$, occurring in Σ_4, can be reduced with the aid of lemma 3.2. We do not write down the result, but just note that - in principle - the sum

$$\sum_{n=1}^{\infty} \check{C}_n(\tau|j)u^n$$

has been determined up to solving the equations for z_j and z'.

Now we turn our attention to the *stationary state* of the original process.

A suggestion for proving limit theorems contained in [SMITH 1958], p.257-258 can be used to prove the existence of $\lim_{n\to\infty} C_n(w|j)$.

The epoch t belongs to a *busy period* if, by definition, the buffer is not empty at time t. In this sense, the busy periods of the original process coincide with those of the modified process. Now if $\rho = \lambda \mathcal{E}\underline{s} < 1$, the termination of a busy period is a certain event in the modified process, hence in the original process. Also, the expected duration of each busy period is finite.

Consider the n-th customer in the original process. The probability that $\underline{w}_n \leq w$ under the condition that he is the k-th of his busy period and that he is of type j, is independent of n. Hence, in order to prove that $C_n(w|j)$ has a limit as $n \to \infty$, it is sufficient to prove that for each k the number

$$P_{n,k} = P\,\{\text{the n-th customer is the k-th of his busy period}\}$$

tends to a limit Q_k, say, with $Q_k \geq 0$ and $\sum_1^\infty Q_k = 1$.

Let $\underline{x}$ be the number of customers in the first busy period. Then it is easy to verify that

$$P_{n,k} = 0 \text{ if } n < k,$$
$$P_{k,k} = P\{\underline{x} \geq k\},$$
$$P_{n,k} = \sum_{i=1}^{n-k} P\{\underline{x} = i\} P_{n-i,k} \text{ if } n > k.$$

Now we need the following lemma. (It is the second part of Theorem 1, Ch. 13 in [FELLER 1967].)

LEMMA 3.4. *Let* $f_1, f_2, \ldots$ *and* $b_0, b_1, \ldots$ *be sequences of non-negative reals with* $f_1 > 0$, $f = \sum_1^\infty f_i = 1$, $g = \sum_i i\, f_i < \infty$, $b = \sum_0^\infty b_i < \infty$. *Let the sequence* $v_0, v_1, \ldots$ *be defined by*

$$v_n = b_n + f_1 v_{n-1} + f_2 v_{n-2} + \ldots + f_n v_0.$$

Then

$$\lim_{n\to\infty} v_n = \frac{b}{g} .$$

REMARK. We have replaced FELLER's requirement that f be non-periodic with the stronger "$f_1 > 0$".

We apply the theorem with $f_i = P\{\underline{x} = i\}$ for all i, and with

$$b_k = \sum_{i=k}^{\infty} f_i,$$

$$b_i = 0 \text{ for } i \neq k.$$

It is easily seen that the conditions of the theorem are satisfied, and that for the generated sequence we have

$$v_0 = \ldots = v_{k-1} = 0,$$

$$v_k = P\{\underline{x} \geq k\},$$

$$v_n = P_{n,k} \text{ for } n > k.$$

Hence we have

$$Q_k = \lim_{n\to\infty} P_{n,k} = \frac{P\{\underline{x}\geq k\}}{\mathcal{E}\underline{x}} .$$

The assertion $\sum_1^{\infty} Q_k = 1$ follows from the fact that $\mathcal{E}\underline{x}$ is finite (and that $\underline{x}$ assumes the values 1,2,... only, so that $\mathcal{E}\underline{x} = \sum_k P\{\underline{x} \geq k\}$).

We may conclude that $C(w|j) = \lim_{n\to\infty} C_n(w|j)$ exists for each j.

From theorem 1 on p.408 of [FELLER 1966] it now follows that the LST

$$\check{C}(\tau|j) = \lim_{n\to\infty} \check{C}_n(\tau|j)$$

exists too, so by applying ABEL's theorem we may conclude that

$$\check{C}(\tau|j) = \lim_{u\to 1} (1-u) \sum_{1}^{\infty} \check{C}_n(\tau|j)u^n = \lim_{u\to 1} (1-u)(\Sigma_3-\Sigma_4).$$

Note that z_j, occurring in Σ_3 and Σ_4, depends on u. However, z_j is a continuous function of u (which can be shown quickly with the aid of a theorem from [FELLER 1966], p.417-418), and from now on we use the notation z_j for the root with smallest absolute value of the "limiting equation"

(3.5) $$z_j = p_j \check{R}_j(\lambda-\lambda z_j).$$

The $\lim_{u\to 1}(1-u)\Sigma_3$ can now be written as

$$\frac{\tau}{\lambda-\tau-\lambda p_j\check{R}_j(\tau)}\left\{\frac{\lambda-\tau}{\tau}\lim_{u\to 1}(1-u)\sum_1^\infty \check{H}_m(\tau)u^m - [\tau := \lambda-\lambda z_j]\right\}.$$

The $\lim_{u\to 1}(1-u)\sum_1^\infty \check{H}_m(\tau)u^m$, occurring here, exists when $\lambda\mathcal{E}\underline{s} < 1$ (see e.g. [TAKÁCS 1962], theorem 10 on p. 69), and is then given by

$$\check{H}(\tau) = \frac{\tau(1-\lambda\mathcal{E}\underline{s})}{\tau-\lambda+\lambda\check{S}(\tau)},$$

which is, of course, the Pollaczek-Khintchine formula mentioned earlier.

Hence, when $\lambda\mathcal{E}\underline{s} < 1$, we have

(3.6) $$\lim_{u\to 1}(1-u)\Sigma_3 = \frac{\tau}{\lambda-\tau-\lambda p_j\check{R}_j(\tau)}\left\{\frac{\lambda-\tau}{\tau}\check{H}(\tau) - [\tau := \lambda-\lambda z_j]\right\}$$

with z_j given by (3.5).

By applying lemma 3.3, we can determine $\lim_{u\to 1}(1-u)\Sigma_4$ in exactly the same manner. The result is

(3.7) $$\lim_{u\to 1}(1-u)\Sigma_4 = \frac{p_j\tau}{\lambda-\tau-\lambda p_j\check{R}_j(\tau)}\left\{\frac{\lambda\check{S}_j(\tau)\check{H}(\tau)}{\tau} - [\tau := \lambda-\lambda z_j]\right\}.$$

So we obtain $\check{C}(\tau\ j)$ as the difference of the right sides of (3.6) and (3.7); after some reduction, we finally obtain

(3.8) $$\check{C}(\tau|j) = \frac{\tau}{\lambda-\tau-\lambda p_j\check{R}_j(\tau)}\left\{\frac{\check{H}(\tau)}{\tau}(\lambda-\tau-\lambda p_j\check{S}_j(\tau)) - [\tau := \lambda-\lambda z_j]\right\}.$$

In the special case of infinite filling rates, we have $\check{R}_j(\tau) \equiv 1$ and $z_j = p_j$, and hence

(3.9) $$\check{C}(\tau|j) = \frac{\tau}{\lambda q_j-\tau}\left\{\frac{\check{H}(\tau)}{\tau}(\lambda-\tau-\lambda p_j\check{S}_j(\tau)) - [\tau := \lambda q_j]\right\}.$$

It is for some purposes more convenient to keep the *conditional* LST $\check{H}_{m+1}(\tau|\underline{j}_m\neq j)$ and its limit as $m \to \infty$ in the formulas. If we use the abbreviation

$$\check{H}(\tau|\underline{j}_{-1}\neq j) = \lim_{m\to\infty}\check{H}_{m+1}(\tau|\underline{j}_m\neq j),$$

then we obtain instead of (3.8) and (3.9):

$$\check{C}(\tau|j) = \frac{\tau q_j}{\lambda-\tau-\lambda p_j \check{R}_j(\tau)} \left\{\frac{\lambda-\tau}{\tau} \check{H}(\tau|\underline{j}_{-1}\neq\underline{j}) - [\tau := \lambda-\lambda z_j]\right\} \tag{3.10}$$

for finite filling-rates, and

$$\check{C}(\tau|j) = \frac{\tau q_j}{\lambda q_j-\tau} \left\{\frac{\lambda-\tau}{\tau} \check{H}(\tau|\underline{j}_{-1}\neq j) - [\tau := \lambda q_j]\right\} \tag{3.11}$$

for infinite filling-rates.

The moments of the limiting distribution may be determined from (3.8) by differentiation and letting $\tau \to 0$. For example, the first moment $\mathcal{E}\underline{c}$ of the unconditional limiting distribution $C(w) = \sum_j p_j C(w|j)$ is given by

$$\mathcal{E}\underline{c} = \mathcal{E}\underline{h} - \sum_j p_j \frac{p_j \mathcal{E}(\underline{s}_j-\underline{r}_j)-T_j}{q_j}, \tag{3.12}$$

where

$$T_j = \frac{\check{H}(\lambda-\lambda z_j)}{\lambda-\lambda z_j} (z_j-p_j\check{S}_j(\lambda-\lambda z_j))$$

and where $\underline{r}_j$, $\underline{s}_j$ and $\underline{h}$ are random variables with LST $\check{R}_j$, $\check{S}_j$, $\check{H}$, respectively.

One might ask whether the moments of the transient distributions converge to the moments of the limiting distribution. A partial answer can be obtained as follows.

Let $\underline{j}_n$ be, as before, the type of the n-th customer, and let $\underline{q}_n$ be the number of customers in the system just after the departure of the n-th customer. Then $\{(\underline{j}_n,\underline{j}_{n+1},\underline{q}_n); n = 1,2,...\}$ is an irreducible, aperiodic Markov-chain with a discrete state space. Let $\underline{f}$ be any non-negative state function. Let ϕ_n be the expected value of $\underline{f}$ just after the n-th transition, and ϕ the expected value of $\underline{f}$ in the stationary state (which is known to exist and to be independent of the initial state). Then, according to theorem 4.3 of [KESTEN & RUNNENBURG 1957]:

$$\lim_{n\to\infty} \phi_n = \phi.$$

In particular, the moments of $\underline{q}_n$ converge to the moments of the queue length $\underline{q}$ in the stationary state.

It has been shown in [LITTLE 1961] and in [JEWELL 1967] that under certain conditions one has

$$\mathscr{E}\underline{c} = \lambda\mathscr{E}\underline{q}.$$

Now JEWELL's Assumption II is easily seen to hold in our case, and his Assumptions I, III, IV follow from $\rho < 1$.

Hence the above question can be answered affirmatively for the first moment of the waiting-time.

3.22 Arbitrarily related filling- and emptying-times

In §3.1 we made the assumption $P\{\underline{r}_n \leq \underline{s}_n\} = 1$, i.e. $r \leq s$ for each realisation (r,s) of $(\underline{r},\underline{s})$. Suppose now that for certain values of j, both $r < s$ and $r > s$ could occur. Then we would have the situation of chapter 2 within a j-sequence. We have refrained from inserting such a complicated process into the process of the present chapter.

Of course, if for certain values of j, each realisation (r,s) satisfies $s \leq r$, so that all mass of $F_j(r,s)$ lies in the first octant, we may "sweep" the mass to the diagonal $r = s$ and thus still apply the results of section 3.21, provided the remaining values of j have all mass in the second octant from the beginning.

3.23 Markov-dependent types

Now we consider a model in which the sequence of types $\underline{j}_1, \underline{j}_2, \ldots$ is an irreducible aperiodic Markov-chain with time-independent transition probabilities p_{ij}. The initial distribution of the types will be denoted by p_{*j}, the stationary distribution by π_j.

Presently, we will have to make a restrictive assumption on the dfs of the emptying times.

Since it seems less natural here to consider the conditional df $C_n(w|j)$, we just take the unconditional df $C_n(w)$. In the same way as in section 3.21 we then find

$$\sum_1^\infty \check{C}_n(\tau)u^n = \Sigma_1 + \Sigma_2$$

where

$$\Sigma_1 = \sum_j \frac{\tau p_{*j} u}{\lambda-\tau-\lambda p_{jj}\check{R}_j(\tau)} \left\{\frac{\lambda-\tau}{\tau} \check{C}_1(\tau|\underline{j}_1=j) - [\tau := \lambda-\lambda z_j]\right\}$$

and

$$\Sigma_2 = \sum_j \sum_{i(\neq j)} \frac{\tau p_{ij} u}{\lambda-\tau-\lambda p_{jj} u\check{R}_j(\tau)} \left\{\frac{\lambda-\tau}{\tau} \sum_1^\infty \check{H}_{m+1}(\tau|\underline{j}_m=i)P\{\underline{j}_m=i\}u^m + \right.$$

$$\left. - [\tau := \lambda-\lambda z_j]\right\}$$

with z_j given by $z_j = p_{jj}u\check{R}_j(\lambda-\lambda z_j)$.

$\check{H}_{m+1}(\tau|\underline{j}_m=i)$ is much more difficult to determine here than in section 3.21, since in the modified process, the service-times are dependent now (through the types).

In order to reduce Σ_2, we now assume that the dfs of the *emptying*-times are independent of the types, as well as the df of $\underline{w}_1$. It then follows that $\check{H}_{m+1}(\tau|\underline{j}_m=i)$ does not depend on i, and we obtain

$$\Sigma_2 = \sum_j \sum_{i(\neq j)} \frac{\tau p_{ij} u}{\lambda-\tau-\lambda p_{jj} u\check{R}_j(\tau)} \left\{\frac{\lambda-\tau}{\tau} \sum_1^\infty \check{H}_{m+1}(\tau)P\{\underline{j}_m=i\}u^m - [\tau := \lambda-\lambda z_j]\right\}.$$

It is still not possible to reduce Σ_2 in the same way as in section 3.21 since $P\{\underline{j}_m=i\}$ does not have a simple form. But we can derive the $\check{C}_n(\tau)$ $(n \geq 2)$ from the formula for Σ_2 as the coefficient of u^n, after which it is easy to obtain $\check{C}(\tau)$.

We first introduce the abbreviation

$$A = \frac{\lambda p_{jj}\check{R}_j(\tau)}{\lambda-\tau}.$$

Since $p_{jj} < 1$, it follows that $|A| < 1$ for sufficiently small values of τ. Now Σ_2 can be written as

$$\Sigma_2 = \sum_j \sum_{i(\neq j)} \frac{u\tau p_{ij}}{\lambda-\tau} \sum_{k=0}^\infty (Au)^k \left\{\frac{\lambda-\tau}{\tau} \sum_1^\infty \check{H}_{m+1}(\tau)P\{\underline{j}_m=i\}u^m - [\tau := \lambda-\lambda z_j]\right\}$$

and the coefficient of u^{n+1} is

$$\sum_j \sum_{i(\neq j)} \frac{\tau p_{ij}}{\lambda-\tau} \sum_{k=0}^{n-1} A^k \left\{\frac{\lambda-\tau}{\tau} \check{H}_{n-k+1}(\tau)P\{\underline{j}_{n-k}=1\} - [\tau := \lambda-\lambda z_j]\right\}.$$

LEMMA 3.5. *If* $B_1, B_2, \ldots$ *is a bounded sequence with the limit* B *and if* $|A| < 1$, *then*

$$\lim_{n\to\infty} \sum_{k=0}^{n-1} A^k B_{n-k} = \frac{B}{1-A} .$$

PROOF. Define $B_0 = B_{-1} = B_{-2} = \ldots = 0$. Then

$$\sum_{k=0}^{n-1} A^k B_{n-k} = \sum_{k=0}^{\infty} A^k B_{n-k},$$

and this has a limit as $n \to \infty$ by the theorem on dominated convergence, and the limit is equal to

$$\sum_{k=0}^{\infty} A^k B = \frac{B}{1-A} . \qquad \square$$

We apply the lemma with $B_k = \check{H}_{k+1}(\tau)P\{\underline{j}_k=1\}$. We take $\rho < 1$ to ensure that $\lim_{k\to\infty} \check{H}_k(\tau)$ exists. The resulting expression for the limit $\check{C}(\tau)$ is

$$\check{C}(\tau) = \sum_j \frac{\tau\pi_j(1-p_{jj})}{\lambda-\tau-\lambda p_{jj}\check{R}_j(\tau)} \left\{\frac{\lambda-\tau}{\tau} \check{H}(\tau) - [\tau := \lambda-\lambda z_j]\right\}$$

with z_j given by $z_j = p_{jj}\check{R}_j(\lambda-\lambda z_j)$.

As a check one may put $\pi_j = p_{jj} = p_j$. This gives

$$\check{C}(\tau) = \sum_j \frac{\tau p_j q_j}{\lambda-\tau-\lambda p_j \check{S}(\tau)} \left\{\frac{\lambda-\tau}{\tau} \check{H}(\tau) - [\tau := \lambda-\lambda z_j]\right\}$$

which is, in the special case where the modified process is independent of the types, equivalent to the unconditional version of (3.10), as is proper.

3.24 The inflow-periods

An inflow-period is defined as a maximal interval during which the amount in the buffer increases. To avoid degenerate cases, we assume that each filling-time is positive with probability 1. A j-inflow-period is defined as an inflow-period with type j in the buffer.

At the arrival of the n-th customer, four cases can be distinguished:

A) The buffer is empty.

B) The buffer is not empty; $\underline{j}_{n-1} \neq \underline{j}_n$.

C) The buffer is not empty; $\underline{j}_{n-1} = \underline{j}_n$; the (n-1)-st customer has completed his filling-operation.

D) The buffer is not empty; $\underline{j}_{n-1} = \underline{j}_n$; the (n-1)-st customer has not completed (perhaps not even started) his filling-operation.

In all cases but the last, an inflow-period starts when the n-th customer begins to fill the buffer. Hence, there are three possibilities for the way in which an inflow-period can start.

The LST $\check{B}_j(\tau)$ of the length of a j-inflow-period under the condition that case A or C occurs and $\underline{j}_n = j$, has been determined by RUNNENBURG. His result is as follows.

THEOREM

$$(3.13) \qquad \check{B}_j(\tau) = \check{G}_j(\lambda-\lambda_j+\tau) + (1-p_j)\,\frac{\check{S}_j(\tau)-\check{G}_j(\lambda-\lambda_j+\tau)}{1-p_j\check{S}_j(\tau)},$$

where $\check{G}_j$ *is the* LST *of the busy periods w.r.t. the emptying-times of the j-customers if the other types are deleted from the system.*

PROOF. Suppose M = 2 and suppose at time 0 a 1-customer enters the empty system. These restrictions are not essential.

We apply a variant of the method of collective marks (cf. [RUNNENBURG 1965]) by introducing a Poisson-process of τ-*catastrophes* with density τ, independent of the process we are studying. The LST $\check{B}_1(\tau)$ can then be interpreted as the probability of the event E_1 = {no τ-catastrophe occurs in the interval $[0, \underline{b}_1]$}, where $\underline{b}_1$ is a random variable with LST $\check{B}_1(\tau)$.

A 2-customer may or may not arrive in the interval $[0,\underline{b}_1]$. The probability of $E_1 \cap$ {no 2-arrival in $[0,\underline{b}_1]$} is equal to $\check{G}_1(\lambda_2+\tau)$, which can be seen at once by interpreting the arrival of a 2-customer as a λ_2-*catastrophe.* The probability of the remaining part of E_1 can be found as follows. Let E_2 be the event {no τ-catastrophe occurs in the service-times of the 1-customers who arrive before the arrival of the first 2-customer}. Let $\underline{k}$ be the number of 1-customers who arrive before the first 2-customer, including the 1-customer who arrives at time 0. Then

$$P\{E_2\} = \sum_{k=1}^{\infty} P\{E_2 \mid \underline{k} = k\}P\{\underline{k} = k\} =$$

$$= \sum_{k=1}^{\infty} \{\check{S}_1(\tau)\}^k p_1^{k-1}(1-p_1) = \frac{(1-p_1)\check{S}_1(\tau)}{1-p_1\check{S}_1(\tau)} .$$

Now the event $E_1 \cap \{$a 2-arrival in $[0,\underline{b}_1]\}$ is part of E_2 and has as its complement in E_2 the event $E_3 = \{$before the first 2-arrival, a 1-inflow period is completed without a τ-catastrophe during its service times; in the other service-times of 1-customer arriving before the first 2-arrival no τ-catastrophe occurs$\}$. Conditioning w.r.t. the number of 1-customers arriving after the completion of the first 1-inflow period and before the 2-arrival, we obtain

$$P\{E_3\} = \check{G}_1(\lambda_2+\tau) \sum_{k=0}^{\infty} \{\check{S}_1(\tau)\}^k p_1^k(1-p_1) = \frac{(1-p_1)\check{G}_1(\lambda_2+\tau)}{1-p_1\check{S}_1(\tau)} ,$$

so that

$$\check{B}_1(\tau) = \check{G}_1(\lambda_2+\tau) + (1-p_1) \frac{\check{S}_1(\tau)-\check{G}_1(\lambda_2+\tau)}{1-p_1\check{S}_1(\tau)}$$

and (3.13) follows by a simple extension. □

A j-inflow period of case B is more difficult to treat, due to the fact that, when such an inflow-period starts, several j-customers will be present in general. Instead, we will only consider a related discrete variable.

Suppose that at the arrival of the n-th customer, case B occurs, and suppose $\underline{j}_n = j$. We will determine the generating function

$$\mathcal{E}(x^{\underline{a}} \mid B;\ \underline{j}_n=j)$$

of the number of j-customers present in the system just after the n-th customer starts to fill, who have arrived earlier than any non-j-customers who may be present at that epoch.

Note that $\{\text{case B}\} = \{\underline{w}_n > 0\} \cap \{\underline{j}_{n-1} \neq \underline{j}_n\}$. Hence, using the abbreviation $E = \{\underline{j}_{n-1} \neq \underline{j}_n = j\}$, we have for $a > 1$:

$$P\{\underline{a}=a \mid B;\ \underline{j}_n=j\} = \frac{P\{\underline{a}=a;\underline{w}_n>0;E\}}{P\{\underline{w}_n>0;E\}} = \frac{P\{\underline{a}=a;E\}}{P\{\underline{w}_n>0;E\}} = \frac{P\{\underline{a}=a \mid E\}}{P\{\underline{w}_n>0 \mid \underline{j}_{n-1}\neq j\}} ,$$

and therefore, after a simple calculation

$$(3.14)\qquad \mathcal{E}(x^{\underline{a}}|B;\ \underline{j}_n=j) = x + \frac{-x+\mathcal{E}(x^{\underline{a}}|E)}{P\{\underline{w}_n>0|\underline{j}_{n-1}\neq j\}}\ .$$

Under the condition E, the variable $\underline{a}^* \overset{\text{def}}{=} \underline{a}-1$ is precisely the number of j-customers who arrive before the first non-j-arrival during the waiting-time $\underline{w}_n$ of the n-th customer, given that he is the first customer of a j-sequence. We start the waiting-time process in the stationary state (cf. §3.21), so that $\underline{w}_n$ has $\check{H}(\tau|\underline{j}_{-1}\neq j)$ as its LST. After a straightforward calculation, it then follows that $\mathcal{E}(x^{\underline{a}}|E)$ is given by

$$(3.15)\qquad \mathcal{E}(x^{\underline{a}}|E) = \frac{x}{1-p_jx}\{1-p_j+p_j(1-x)\check{H}(\lambda-\lambda_jx|\underline{j}_{-1}\neq j\}.$$

It remains to determine the denominator of the right-hand-side of (3.14). By letting $\tau\to\infty$ in lemma 3.3, we find $P\{\underline{w}_n>0|\underline{j}_{n-1}=j\} = 1-\check{S}_j(\lambda)\check{W}(\lambda)$ and hence

$$P\{\underline{w}_n>0|\underline{j}_{n-1}\neq j\} = 1 - \frac{\check{S}(\lambda)-p_j\check{S}_j(\lambda)}{1-p_j}\check{W}(\lambda).$$

In constrast to the situation in §3.21, there is no simple way here to eliminate the condition E either from (3.14) or from (3.15) due to the fact that the probability of an arbitrary inflow-period being of type j, is not equal to p_j. E.g., consider a very congested system with M = 2. Then the above-mentioned probability is approximately $\frac{1}{2}$ (j=1,2), since one will observe an alternation of 1- and 2-inflow-periods most of the time.

3.3 FILLING WITH PRIORITIES

We conclude this chapter with some comments on a model in which the filling-operation is subject to priorities.

We recall that the emptying discipline in the present chapter is "the emptying line is busy whenever the buffer contains a positive amount". Since there is only one buffer, such an emptying discipline is, in a sense, the only reasonable choice. Hence, here we do not have the difficulty alluded to in §2.4.

For the filling discipline there are many possibilities. Although linear priorities are the most common in the literature, we have hardly considered that possibility; most of our effort has been spent on a model with M = 2 in which the filling operation is subject to alternating priorities as defined in §2.3. However, the results are quite complicated and of an implicite nature. Moreover, we have not been able so far to surmount certain technical difficulties.

After this negative information, it is perhaps desirable to motivate our choice of alternating priorities. In the first place, from the point of view of efficiency, alternating priorities seem to be tailor-made for a buffer model of the type in this chapter. Secondly, they are mathematically attractive because of the fact that, as long as customers of type 1 are served, the customers of type 2 can be neglected entirely, and vice versa. (Cf. the end of §2.3 and also §4.34.)

CHAPTER 4

K INFINITE BUFFERS; $2 \leq K < M$

4.1 INTRODUCTION

In this chapter, we assume K infinite buffers which are almost empty, and M types of customers with $2 \leq K < M$.

The n-th customer arrives at time $\underline{y}_0 + \ldots \underline{y}_{n-1}$ $(n \geq 1)$ where each $\underline{y}_i$ has an exponential distribution with $\mathcal{E}\underline{y}_i = \lambda^{-1}$ and where $\underline{y}_0, \underline{y}_1, \ldots$ are independent.

The type of the n-th customer is denoted by $\underline{j}_n$. We assume that p_j, defined by

$$p_j = P\{\underline{j}_n = j\} \qquad (j = 1,\ldots,M)$$

does not depend on n, that all $p_j > 0$, and that the variables $\underline{j}_1, \underline{j}_2, \ldots$ are mutually independent and independent of the arrivals process.

Unless stated otherwise, the filling-times are identically 0. The emptying-time for the n-th customer is denoted by $\underline{s}_n$; we assume that $\underline{s}_1, \underline{s}_2, \ldots$ are independent, independent of the arrivals process, and that each $\underline{s}_n$ is independent of $\underline{j}_1, \underline{j}_2, \ldots$ except (possibly) $\underline{j}_n$.

The waiting-time of the n-th customer is denoted by $\underline{w}_n$. We assume that $\underline{w}_1$ has a given distribution, which may depend on $\underline{y}_0$ and $\underline{j}_1$, but which is independent of the rest of the process.

Because of the complexity of the model, there is no obvious choice for the filling discipline.

The very popular 'fifo' discipline has several points in its favour. In the first place, it is considered *fair* in many situations where the customers have diverging interests. Secondly, it is *optimal* (it minimizes $\sigma^2 \underline{w}$) within a class of disciplines satisfying certain assumptions; see [Kingman 1962] or [Cohen 1969], p.463 ff. However, in setting up a mathematical model, these advantages have little weight. What really turns the scale is the fact that the 'fifo' discipline is mathematically *manageable*.

Now, in the complex models we are considering, the 'fifo' discipline may be far from optimal and far from simple. The actual strategy chosen in real situations of this type will in general depend on various details, that we have not incorporated in our framework. What we will do in the sequel is to consider certain restrictions or special cases which are at least manageable and perhaps realistic.

4.2 FIXED ASSIGNMENT OF BUFFERS TO TYPES

In this section, we restrict the possible strategies by partitioning the types into K groups and assigning one buffer to each group. A customer may fill only the buffer of his own group.

We number the buffers $1,2,\ldots,K$. The *number* of types assigned to buffer i be m_i ($m_i > 0$, $\sum m_i = M$).

The restriction having been made, a fairly natural choice for the filling and emptying disciplines presents itself, depending on the values of $m_1,\ldots,m_K$.

4.21 All groups but one of size 1

Suppose $m_1 = \ldots = m_{K-1} = 1$ and hence $m_K = M - K+1$. The emptying operator gives preemptive priority to buffer K. Due to the nature of our problem, the preemptive priority is of the resume type here. (This means the following: suppose the emptying operator has interrupted the service of a load, L say, of a type $< K$, due to the arrival of a type $\geq K$ in buffer K. Then the emptying operator will after some time *resume* his service of load L, and the time spent earlier on load L is not lost.) The buffers $1,\ldots,K-1$ are emptied in any order. The filling-discipline is as follows: customers of types $1,\ldots,K-1$ fill their buffer at their arrival (they never have to wait). The customers of types $K,\ldots,M$ (who share buffer K) form a queue and fill buffer K in order of arrival (under the usual restriction that a buffer may contain at most one type).

The process of the $K,\ldots,M$-customers is not at all influenced by the other types and is given by the results of section 3.2 (with filling-times 0).

Even when the filling- and emptying-times for types $K,\ldots,M$ have a simultaneous distribution "in the second octant", the results of § 3.2 apply.

However, when there is a type $\geq K$ with a sufficiently large filling-time (larger than its emptying-time), the process may not become stationary even though it may be possible that a process with a stationary state is obtained by giving low priority to such a type at the emptying-line, causing this line to operate at full capacity when the system is not empty.

If there are one or more types < K with non-zero filling-times, the situation is essentially more complex. We do not consider this case.

Returning to the special case of *all* filling-times equal to zero, we note that in such a case one can, in principle, choose an optimal assignment of the types to the buffers from the $\binom{M}{K-1}$ possible assignments, where "optimal" is meant in the sense of giving lowest expected waiting-time of an arbitrary customer in the stationary state.

4.22 All $m_i \geq 2$

Suppose that all $m_i \geq 2$, i.e. each buffer is shared by at least two types. The empyting takes place in order of arrival. The filling-discipline is as follows: within each group, the customers fill their buffer in order of arrival, and otherwise as soon as possible. Note that the filling does not, as a rule, take place in order of arrival.

We consider, along with the original process, a modified process in which each customer who finds the emptying-line occupied, postpones his filling until the total amount of goods in the buffers becomes 0.

Let $\underline{J}_n$ be the group of the n-th customer, $\pi_J = P\{\underline{J}_n=J\}$ $(J = 1,\ldots,K)$. Let $\underline{a}_n$ be the arrival time of the n-th customer, $\underline{c}_n$ the time at which he fills the buffer in the original process, and $\underline{e}_n$ the time at which the emptying-line completes the operation on the n-th load.

Note that the epoch $\underline{a}_n$ coincides for the original and the modified process, as well as the epoch $\underline{e}_n$.

Consider the n-th customer. Suppose $\underline{j}_n = j$ and suppose type j belongs to group J.

The number $\underline{t}_1$ (depending on n) is defined as follows. If there exists an index $t < n$ with $\underline{J}_t = J$, then $\underline{t}_1$ is the largest such index. If $\underline{J}_t \neq J$ for all $t < n$, then $\underline{t}_1 = 0$ by definition. In the latter case, we also assign the artificial value 0 to $\underline{j}_0, \underline{e}_0, \underline{s}_0$ and $\underline{w}_0$.

We now consider the waiting-time of the n-th customer in the original process, i.e. $\underline{w}_n = \underline{c}_n - \underline{a}_n$. There are two cases:

1. $\underline{j}_{\underline{t}_1} = j$,

2. $\underline{j}_{\underline{t}_1} \neq j$.

In case 1, $\underline{c}_n \geq \underline{a}_n$ and $\underline{c}_n \geq \underline{c}_{\underline{t}_1}$, and since there are no further restrictions on $\underline{c}_n$, we have

$$\underline{c}_n = \max(\underline{a}_n, \underline{c}_{\underline{t}_1}).$$

It follows that in case 1, $\underline{w}_n$ satisfies the relation

(4.1) $$\underline{w}_n = \max\{0, \underline{w}_{\underline{t}_1} - (\underline{y}_{\underline{t}_1} + \ldots + \underline{y}_{n-1})\}.$$

In case 2 (which includes the artificial case), $\underline{c}_n \geq \underline{a}_n$ and $\underline{c}_n \geq \underline{e}_{\underline{t}_1}$, hence

$$\underline{c}_n = \max(\underline{a}_n, \underline{e}_{\underline{t}_1})$$

and

(4.2) $$\underline{w}_n = \max\{0, \underline{g}_{\underline{t}_1} + \underline{s}_{\underline{t}_1} - (\underline{y}_{\underline{t}_1} + \ldots + \underline{y}_{n-1})\},$$

where $\underline{g}_{\underline{t}_1}$ is the waiting-time of customer $\underline{t}_1$ in the modified process, given that $\underline{j}_n = j$, $\underline{j}_{\underline{t}_1} \neq j$, $\underline{J}_{\underline{t}_1} = J$.

Now (4.1) can be used to express $\underline{w}_n$ in terms of a waiting-time in the modified process. In other words, when case 1 applies, we may continue backwards, neglecting non-J customers, until a J-non-j customer occurs. More precisely: we inductively define a strictly decreasing sequence of indices $\underline{t}_1, \underline{t}_2, \ldots$, all depending on n, as follows. Suppose $i \geq 2$, $\underline{t}_{i-1} > 0$ and $\underline{j}_{\underline{t}_{i-1}} = j$. If there exists an index $t < \underline{t}_{i-1}$ with $\underline{J}_t = J$ then $\underline{t}_i$ is the largest such index. If $\underline{J}_t \neq J$ for all $t < \underline{t}_{i-1}$, then $\underline{t}_i = 0$ by definition. The sequence $\underline{t}_1, \underline{t}_2, \ldots$ has a finite number of terms: when $\underline{t}_i = 0$ or $\underline{j}_{\underline{t}_i} \neq j$, no further $\underline{t}$'s are defined. The length of the sequence is called $\underline{r}$. By applying (4.1) repeatedly, it is then easily seen that $\underline{w}_n$ can be written as follows.

(4.3) $$\underline{w}_n = \max\{0, \underline{g}_{\underline{t}_{\underline{r}}} + \underline{s}_{\underline{t}_{\underline{r}}} - (\underline{y}_{\underline{t}_{\underline{r}}} + \ldots + \underline{y}_{n-1})\}.$$

We are now ready to determine the limit as $n \to \infty$ of the conditional LST of $\underline{w}_n$ under the condition $\underline{j}_n = j$. We will usually abbreviate an event like $\{\underline{j}_n=j,\underline{t}_1=t_1\}$ to $\{j,t_1\}$, etc.

$$(4.4)\qquad \mathcal{E}(e^{-\tau\underline{w}_n}|\underline{j}_n=j) =$$

$$\sum_r \sum_{t_1,\dots,t_r} \mathcal{E}(e^{-\tau\underline{w}_n}|j,r,t_1,\dots,t_r)P\{r,t_1,\dots,t_r|j\} =$$

$$\sum_r \sum_{t_1,\dots,t_r} \mathcal{E}(e^{-\tau \max\{0,\underline{g}_{t_r}+\underline{s}_{t_r}-(\underline{y}_{t_r}+\dots+\underline{y}_{n-1})\}}|j,r,t_r) \cdot$$

$$\cdot\ P\{r,t_1,\dots,t_r|j\}$$

It is easily shown that

$$(4.5)\qquad P\{r,t_1,\dots,t_r|j\} = \begin{cases} (1-\pi_J)^{n-r}p_j^{r-1} & \text{if } t_r = 0, \\ (1-\pi_J)^{n-t_r-r}p_j^{r-1}(\pi_J-p_j) & \text{if } t_r > 0, \end{cases}$$

hence the summand in (4.4) is independent of $t_1,\dots,t_{r-1}$, and the summation over $t_1,\dots,t_{r-1}$ may be executed, giving

$$(4.6)\qquad \mathcal{E}(e^{-\tau\underline{w}_n}|j) = \sum_{r=1}^{n} \sum_{t_r=0}^{n-r} \binom{n-t_r-1}{r-1} \cdot P\{r,t_1,\dots,t_r|j\} \cdot$$

$$\cdot\ \mathcal{E}(e^{-\tau\{\max 0,\underline{g}_{t_r}+\underline{s}_{t_r}-(\underline{y}_{t_r}+\dots+\underline{y}_{n-1})\}}|j,r,t_r).$$

For the moment, we forget about the interpretation of the variables $\underline{y}_i$; we introduce variables $\underline{y}_{-1},\underline{y}_{-2},\dots$ each of which is exponentially distributed with parameter λ, and such that the set of variables $\underline{y}_{-1},\underline{y}_{-2},\dots$ is independent of the rest of the process.

We claim that $\mathcal{E}(e^{-\tau\underline{w}_n}|j)$ may be approximated by

$$(4.7)\qquad S_n = \sum_{r=1}^{n} \sum_{t_r=-\infty}^{n-r} \binom{n-t_r-1}{r-1} \mathcal{E}(e^{-\tau \max\{0,\underline{g}_{t_r}+\underline{s}_{t_r}-(\underline{y}_{t_r}+\dots+\underline{y}_{n-1})\}}|j,t_r,r).$$

$$\cdot\ (1-\pi_J)^{n-t_r-r}p_j^{r-1}(\pi_J-p_j).$$

Indeed, since the absolute value of every LST is at most 1, and since for each r, the terms with $t_r > 0$ are the same in both sums of (4.6) and (4.7), we have

$$(4.8) \qquad \left|\mathcal{E}(e^{-\tau \underline{w}_n}|j)-S_n\right| \le \sum_{r=1}^{n} \binom{n-1}{r-1}(1-\pi_J)^{n-r}p_j^{r-1} + \\ + \sum_{r=1}^{n}\sum_{t_2=-\infty}^{0}\binom{n-t_r-1}{r-1}(1-\pi_J)^{n-t_r-r}p_j^{r-1}(\pi_J-p_j),$$

and after substituting $r = s+1$ in both sums and $n-t_r = t$ in the second sum, we readily find that the right-hand-side of (4.8) may be reduced to

$$(1-\pi_J+p_j)^{n-1} + (1-\pi_J+p_j)^n,$$

which tends to 0 as $n \to \infty$. (Here we make use of the fact that each group contains at least two types).

Hence

$$\lim_{n\to\infty}\left|\mathcal{E}(e^{-\tau \underline{w}_n}|j)-S_n\right| = 0.$$

Now we assume that the expected (unconditional) emptying-time is less than λ^{-1} (so that the modified process has a stationary state) and that the modified process is started in the stationary state.

Then the distribution of $\underline{g}_{t_r}$ under the condition $\{j,t_r,r\}$ does not depend on t_r nor on r. If we introduce variables $\underline{g}^*$ and $\underline{s}^*$, denoting the waiting-time in the stationary modified process and the emptying-time respectively, for an arbitrary j-customer, then (4.7) may be written as follows:

$$S_n = \sum_{r=1}^{n}\sum_{t_r=-\infty}^{n-r}\binom{n-t_r-1}{r-1}\mathcal{E}e^{-\tau\max\{0,\underline{g}^*+\underline{s}^*-\underline{y}_{t_r}-\cdots-\underline{y}_{n-1}\}} \times \\ \times (1-\pi_J)^{n-t_r-r}p_j^{r-1}(\pi_J-p_j),$$

or with $t = n-t_r$ and $s = r-1$:

$$S_n = \sum_{s=0}^{n-1}\sum_{t=s+1}^{\infty}\binom{t-1}{s}\mathcal{E}e^{-\tau\max\{0,\underline{g}^*+\underline{s}^*-\underline{y}_{n-t}-\cdots-\underline{y}_{n-1}\}} \times \\ \times (1-\pi_J)^{t-s-1}p_j^s(\pi_J-p_j) = \\ = \sum_{s=0}^{n-1}\sum_{t=s+1}^{\infty}\binom{t-1}{s}\mathcal{E}e^{-\tau\max\{0,\underline{g}^*+\underline{s}^*-\underline{y}_1-\cdots-\underline{y}_t\}}(1-\pi_J)^{t-s-1}p_j^s(\pi_J-p_j),$$

which is a sum of n *positive* terms, each of which is independent of n. Since S_n is moreover bounded from above by 1, it follows that $\lim_{n\to\infty} S_n = S$ exists. Hence

$$\check{W}(\tau|j) \overset{\text{def}}{=} \lim_{n\to\infty} \mathcal{E}(e^{-\tau \underline{w}_n}|j)$$

exists too, and is equal to S.

Now it is easily shown that the form

$$\sum_{s=0}^{n-1} \sum_{t=s+1}^{\infty} \binom{t-1}{s} \mathcal{E} e^{-\tau(\underline{y}_1+\ldots+\underline{y}_t)} (1-\pi_J)^{t-s-1} p_j^s (\pi_J - p_j)$$

may be reduced to $\frac{\lambda(\pi_J-p_j)}{\tau+\lambda(\pi_J-p_j)} \left\{1 - \left(\frac{\lambda p_j}{\tau+\lambda\pi_J}\right)^n\right\}$, which means that $\check{W}(\tau|j)$ (or S) may be written as

$$(4.9) \qquad \check{W}(\tau|j) = \mathcal{E}(e^{-\tau \max(0,\underline{g}^*+\underline{s}^*-\underline{z})}|j)$$

where $\underline{z}$ is exponential with parameter $\lambda(\pi_J-p_j)$ and independent of $\underline{g}^*$ and $\underline{s}^*$. Hence $\check{W}(\tau|j)$ is just the conditional LST of a customer in an M|G|1 model (with $\lambda^{-1}(\pi_J-p_j)^{-1}$ as the expected arrival interval), under the condition that the predecessor of that customer is a J-non-j customer.

Hence, we may apply lemma 3.3 with λ replaced by $\lambda(\pi_J-p_j)$, $\check{S}_j(\tau)$ by

$$\check{S}(\tau|J\text{-non-}j) \overset{\text{def}}{=} \mathcal{E}(e^{-\tau\underline{s}_n}|\underline{J}_n=J,\underline{j}_n\neq j),$$

and $\check{H}(\tau)$ by the unconditional LST of an arbitrary customer's waiting-time in the stationary state of the modified process. The result is

$$\check{W}(\tau|j) = \frac{\lambda(\pi_J-p_j)\tau}{\lambda(\pi_J-p_j)-\tau} \left\{\frac{\check{S}(\tau|J\text{-non-}j)\check{H}(\tau)}{\tau} - [\tau := \lambda(\pi_J-p_j)]\right\}$$

4.23 General m_i

If at least two but not all m_i are greater than 1, we speak of "general m_i". Suppose, without restriction, that $m_1 = \ldots = m_L = 1$; $m_{L+1},\ldots,m_K > 1$; $1 \le L \le K-2$. A suitable choice of the filling- and emptying disciplines will enable us to treat the case of general m_i by combining sections 4.21 and 4.22. In fact, types 1,...,L are given preemptive priority at the filling-line, whereas they have access to the emptying-line only when there are no loads of types L+1,...,M in the system. Types L+1,...,M are subject to the discipline of §4.22.

As before, the waiting-time of types 1,...,L is 0, while the waiting-times of types L+1,...,M are not influenced by types 1,...,L.

Note that this policy, although somewhat complicated to formulate, is mathematically surprisingly manageable and also has the appearance of making a very efficient use of the buffers, given the fixed assignment.

Also note that section 4.21 is not a special case of the present case, due to the fact that it is essential that the π_J for types L+1,...,M are strictly between 0 and 1.

4.3 TWO RELAXATIONS OF A RESTRICTED PROCESS

In this section and its subsections we take $K = 2$ unless stated otherwise. (In any case we keep $M > K \geq 2$). The emptying discipline is always: "Empty in order of arrival". The filling discipline is defined in the subsections. Since filling occurs at an infinite rate, the emptying operation governs the existence of a stationary state, and a necessary condition for this to exist is $\lambda \mathcal{E}\underline{s} < 1$.

4.31 The restricted process

Suppose filling takes place in order of arrival under the restriction that a total of at most two loads is allowed in the two buffers together. (The easiest way to visualize this restriction is to imagine that the odd-numbered customers use buffers 1, the even-numbered customers buffer 2. But we do not require the customers to do this. The reason is given in §4.32). The resulting process is called the *restricted process*.

Of course, when the above restriction is not present in a practical situation, it would be unwise to impose this discipline. But it is fairly easy to conceive of practical situations where the restriction does occur. (See also the interpretation at the end of this subsection.) As suggested by the titles of §4.3 and §4.31, the main reason why we consider the restricted process is that we need the results in §4.32 and 4.33.

We are interested in $\underline{h}_n$, the waiting-time of the n-th customer. Let us define the following epochs pertaining to the n-th customer:

$\underline{a}_n$: arrival,
$\underline{b}_n$: filling,
$\underline{c}_n$: start of emptying,
$\underline{d}_n$: end of emptying,

so that $\underline{h}_n = \underline{b}_n - \underline{a}_n$, $\underline{y}_n = \underline{a}_{n+1} - \underline{a}_n$, $\underline{s}_n = \underline{d}_n - \underline{c}_n$.

Further we define

$$\underline{e}_n = \underline{c}_n - \underline{b}_n,$$
$$\underline{w}_n = \underline{c}_n - \underline{a}_n.$$

Figure 4.1 gives a pictorial summary of these definitions.

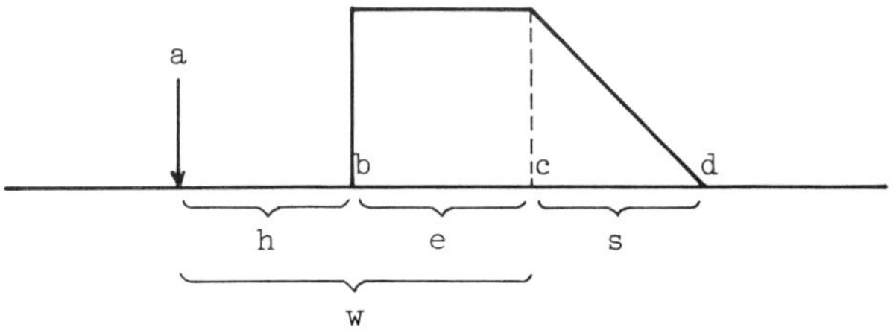

Figure 4.1

THEOREM 4.1. *In the model and the notation introduced above we have*

$$\underline{w}_{n+1} = \max(0, \underline{w}_n + \underline{s}_n - \underline{y}_n), \tag{4.10}$$

$$\underline{h}_{n+1} = \max(0, \underline{w}_n - \underline{y}_n), \tag{4.11}$$

and $\underline{w}_n$, $\underline{s}_n$ *and* $\underline{y}_n$ *are independent.*

PROOF. From the definition of the process we see that

$$\underline{c}_{n+1} = \max(\underline{a}_{n+1}, \underline{d}_n)$$

hence

$$\underline{w}_{n+1} = \underline{c}_{n+1}-\underline{a}_{n+1}$$
$$= \max(\underline{a}_{n+1},\underline{d}_n) - \underline{a}_{n+1}$$
$$= \max(0,\underline{c}_n+\underline{s}_n-\underline{a}_n-\underline{y}_n)$$
$$= \max(0,\underline{w}_n+\underline{s}_n-\underline{y}_n)$$

which proves (4.10).

We also have

$$\underline{b}_{n+1} = \max(\underline{a}_{n+1},\underline{d}_{n-1}),$$

since the (n+1)-st customer fills the buffer either at his arrival or at the epoch at which the load of the (n-1)-st customer leaves the buffer, whichever occurs last. Hence $\underline{b}_{n+1} \geq \max(\underline{a}_n,\underline{d}_{n-1}) = \underline{c}_n$ and $\underline{b}_{n+1} \geq \underline{a}_{n+1}$ and therefore

(4.12) $$\underline{b}_{n+1} \geq \max(\underline{c}_n,\underline{a}_{n+1}).$$

On the other hand, from $\underline{c}_n = \max(\underline{a}_n,\underline{d}_{n-1})$ it follows that $\underline{c}_n \geq \underline{d}_{n-1}$, so that

(4.13) $$\max(\underline{a}_{n+1},\underline{c}_n) \geq \max(\underline{a}_{n+1},\underline{d}_{n-1}) = \underline{b}_{n+1}.$$

From (4.12) and (4.13) we infer

$$\underline{b}_{n+1} = \max(\underline{a}_{n+1},\underline{c}_n),$$

hence

$$\underline{h}_{n+1} = \underline{b}_{n+1}-\underline{a}_{n+1} = \max(0,\underline{c}_n-\underline{a}_n-\underline{y}_n)$$
$$= \max(0,\underline{w}_n-\underline{y}_n).$$

The independence of $\underline{w}_n$, $\underline{s}_n$, $\underline{y}_n$ is obvious. □

We have not made use of the fact that $\underline{y}_0,\underline{y}_1,\underline{y}_2,\ldots$ are exponential.

From theorem 4.1 we see that the results of Chapter 3 can be applied to the variables $\underline{w}_n$, while (4.11), or more conveniently the following result, enables us to obtain $\underline{h}_n$ from $\underline{w}_n$.

LEMMA 4.1. *For* $\mathrm{Re}(\tau) \geq 0$, $\tau \neq \lambda$, *the* LST $\check{H}_{n+1}$ *of* $\underline{h}_{n+1}$ *is given by*

$$(4.14) \qquad \check{H}_{n+1}(\tau) = \frac{\lambda\check{W}_n(\tau) - \tau\check{W}_n(\lambda)}{\lambda-\tau} .$$

PROOF.

$$\mathcal{E}e^{-\tau\underline{h}_{n+1}} = \int_{w=0}^{\infty} dW_n(w) \left\{\int_{y=0}^{w} e^{-\tau(w-y)}\lambda e^{-\lambda y}dy + \int_{y=w}^{\infty} \lambda e^{-\lambda y}dy\right\},$$

which reduces to 4.14. □

Lemma 4.1 and lemma 3.1 may now be used to obtain a closed form for the generating function of $\check{H}_n(\tau)$.

If we assume $\lambda\mathcal{E}\underline{s} < 1$, the existence of $\check{W}(\tau) = \lim_{n\to\infty} \check{W}_n(\tau)$ follows from lemma 3.3. From (4.14) we then see at once that

$$\check{H}(\tau) = \lim_{n\to\infty} \check{H}_n(\tau)$$

exists too, and is given by

$$(4.15) \qquad \check{H}(\tau) = \frac{\lambda\check{W}(\tau) - \tau\check{W}(\lambda)}{\lambda-\tau} .$$

By differentiation, we obtain the following simple formula for the first moment:

$$\mathcal{E}\underline{h} = \mathcal{E}\underline{w} - \frac{1-\check{W}(\lambda)}{\lambda} .$$

The above results may be extended to $K \geq 2$, although much of the simplicity is lost.

In concluding this subsection, we note that the variable $\underline{h}_n$ has a simple interpretation in an ordinary 1-server model: it is the amount of time during which a customer has to wait (from his arrival) until the first epoch at which the number in the queue before him, including the customer who is being served, is at most K-1.

4.32 First relaxation

In this subsection we consider a discipline which may best be defined with the aid of the process of § 4.31 . Let $\underline{j}_n$ be the type of the n-th customer. Then we advance the filling epoch from $\underline{b}_n$ to $\underline{x}_n$, where $\underline{x}_n$ $(n \geq 1)$ is defined by

$$\underline{x}_n = \begin{cases} \max(\underline{a}_n, \underline{x}_{n-1}) \text{ if } \underline{j}_n = \underline{j}_{n-1} \text{ and } n \geq 2, \\ \\ \underline{b}_n \text{ otherwise.} \end{cases}$$

This means that a customer advances his filling-epoch provided he is of the same type as his predecessor. Also, we now require that such a customer uses the <u>same</u> buffer as his predecessor. Here we see why in §4.31 the customers are not required to use the buffers alternatingly. In fact, if the customers would do this in the present process, a buffer might contain two types. For example, suppose a 1-customer enters an empty system and fills buffer 1, and suppose that within his emptying-time, a 2-customer arrives, immediately followed by another 2-customer. It is easy to see that in this case it is not possible to advance the filling-epoch of the third customer to the epoch $\underline{x}_n$ as defined above.

The resulting process is called the *first relaxation.*

We claim that the above definition is permitted in the sense that the condition "the number of types in the buffers is at most 2" is always satisfied.

It is convenient to state and prove this together with some simple auxiliary relations.

<u>THEOREM 4.2</u>.

A) $\underline{a}_n \leq \underline{x}_n \leq \underline{b}_n$ $(n \geq 1)$,

B) $\underline{x}_1 \leq \underline{x}_2 \leq \ldots$,

C) *At any time, each buffer contains at most one type.*

PROOF.

A. By definition $\underline{x} = \underline{b}_1$. Suppose $\underline{a}_{n-1} \le \underline{x}_{n-1} \le \underline{b}_{n-1}$ $(n \ge 2)$. Then either $\underline{x}_n = \underline{b}_n$ or $\underline{a}_n \le \underline{x}_n = \max(\underline{a}_n, \underline{x}_{n-1}) \le \max(\underline{a}_n, \underline{b}_{n-1}) \le$ $\le \max(\underline{a}_n, \underline{b}_n) = \underline{b}_n$.

B. If $\underline{x}_n = \max(\underline{a}_n, \underline{x}_{n-1})$, then $\underline{x}_n \ge \underline{x}_{n-1}$ is immediate. If $\underline{x}_n = \underline{b}_n$, then $\underline{x}_n = \underline{b}_n \ge \underline{b}_{n-1} \ge \underline{x}_{n-1}$.

C. Consider a fixed realization of the process of §4.31. We successively determine $x_1, x_2, \ldots$. We claim that before the shift of the filling epoch of customer n from b_n to x_n, type j_n is present in the system at each epoch of the interval (x_n, b_n). Assuming that the claim is justified with respect to customers 1,2,...,n-1, we show that it is justified for customer n.

If $j_{n-1} \neq j_n$ or $a_n = b_n$, then no shift takes place and there is nothing to prove.

Suppose $j_n = j_{n-1}$ and $a_n < b_n$. The shift takes place along the interval $(\max(a_n, x_{n-1}), b_n) \subseteq (x_{n-1}, b_n) = (x_{n-1}, b_{n-1}) \cup [b_{n-1}, b_n)$. Now (x_{n-1}, b_{n-1}) contains type j_n $(=j_{n-1})$ because of the induction hypothesis. And $a_n < b_n$ implies $b_n = d_{n-2} < d_{n-1}$ hence $[b_{n-1}, b_n) \subset [b_{n-1}, d_{n-1})$ during which interval the system obviously contains j_{n-1} $(=j_n)$, so the shift is permitted.

To complete the proof, we note that since $x_1 = b_1$, the induction can be started. □

A practical situation in which the above discipline is effectuated may arise when for certain reasons (e.g. technical or administrative)

a) the filling-line is switched to the other buffer if and only if a customer arrives who differs in type from his predecessor,

b) such a customer has to wait until the buffer is empty.

Now suppose we are interested in $\check{G}(\tau|j)$, the limiting LST of the waiting-time $\underline{g}_n = \underline{x}_n - \underline{a}_n$ of the n-th customer under the condition $\underline{j}_n = j$.

Within a sequence of consecutive customers of the same type, we have the situation of §3.21 (with infinite filling-rates). In fact, it can be shown that formula (3.11) continues to hold. In the present process, the situation is actually a little simpler, since $\check{H}(\tau|\underline{j}_{-1}\neq j)$, occurring in (3.11), does not depend on j here. This can be shown as follows.

Consider the (n+1)-st customer in the restricted process of §4.31. His waiting-time $\underline{h}_{n+1}$ is given by (4.11), i.e.

$$\underline{h}_{n+1} = \max(0, \underline{w}_n - \underline{y}_n)$$

and neither $\underline{w}_n$ nor $\underline{y}_n$ depends on $\underline{j}_n$. Hence

$$\check{H}(\tau|\underline{j}_{-1}\neq j) = H(\tau),$$

and we have

$$\check{G}(\tau|j) = \frac{\tau q_j}{\lambda q_j - \tau}\left\{\frac{\lambda-\tau}{\tau}\check{H}(\tau) - [\tau := \lambda q_j]\right\}$$

with $\check{H}(\tau)$ given by (4.15).

If K > 2, this is still true for the present process, in the sense that (3.11) holds with a different, more complicated expression for $\check{H}(\tau)$.

4.33 Second relaxation

Let us define $\underline{a}_n$ and $\underline{b}_n$ as in §4.31, and $\underline{x}_n$ $(n \geq 3)$ now by

$$(4.16) \qquad \underline{x}_n = \begin{cases} \max(\underline{a}_n, \underline{x}_{n-1}) & \text{if } \underline{j}_{n-1} = \underline{j}_n, \\ \max(\underline{a}_n, \underline{x}_{n-2}) & \text{if } \underline{j}_{n-1} \neq \underline{j}_n = \underline{j}_{n-2}, \\ \underline{b}_n & \text{otherwise.} \end{cases}$$

Further we define: $\underline{x}_1 = \underline{b}_1$; $\underline{x}_2 = \underline{b}_1$ if $\underline{j}_1 = \underline{j}_2$, $\underline{x}_2 = \underline{b}_2$ if $\underline{j}_1 \neq \underline{j}_2$. As in §4.32 we shift the filling epochs from $\underline{b}_n$ to $\underline{x}_n$. We also require the customers to fill the buffer which already contains their type, if possible.

THEOREM 4.3.

A) $\underline{a}_n \leq \underline{x}_n \leq \underline{b}_n$

B) *Filling does not take place in order of arrival.*

C) *At any time, each buffer contains at most one type.*

PROOF.

A. Analogous to part A of theorem 4.2.

B. It suffices to give an example occurring with positive probability starting from an empty system. Suppose a 1-customer arrives, and then within his emptying-time a 2-customer, a 3-customer, and a 2-customer, respectively. Then the fourth customer fills before the third, as is easily seen.

C. We start in the same manner as in the proof of theorem 4.2, part C. Consider a fixed realization. If $j_{n-1} \neq j_n \neq j_{n-2}$ or $a_n = b_n$, there is nothing to prove. If $j_n = j_{n-1}$ and $a_n < b_n$, the proof of §4.32 can be copied. We only have to consider the case: $j_{n-1} \neq j_n = j_{n-2}$ and $a_n < b_n$.

From $a_n < b_n$ we see that $b_n = d_{n-2}$. The shift takes place along the interval $(x_n, b_n) = (\max(a_n, x_{n-2}), b_n) \subseteq (x_{n-2}, d_{n-2}) = (x_{n-2}, b_{n-2}) \cup [b_{n-2}, d_{n-2})$, and in each of the last two intervals the system contains type $j_{n-2} = j_n$, so the shift is again permitted.

To complete the proof, we note that from the definition of x_1 en x_2 it is obvious that the induction can be started. □

REMARK. The above proof is comparatively simple because of the use of the quantities d_n. The process is in fact quite complicated. For example, it may happen that during an interval of the form (x_n, b_n) all types occur in the buffers. Also, two intervals (x_{n_1}, b_{n_1}) and (x_{n_2}, b_{n_2}) with $j_{n_1} \neq j_{n_2}$ may have an intersection of positive length.

Practical situations where this discipline may occur are of the same kind as those of §4.32, the difference being that here it requires a *pair* of non-j-customers to interrupt a j-sequence.

We will now determine $\check{G}_n(\tau|j)$, that is: the conditional LST of $\underline{g}_n = \underline{x}_n - \underline{a}_n$ under the condition $\underline{j}_n = j$. First we require some definitions.

From the n-th customer counting backward, let $\underline{m}_n$ be the number of predecessors until the occurrence of two successive non-j-customers. More precisely: $\underline{m}_1 = 1$, $\underline{m}_2 = 2$, and for $n \geq 3$:

$$\underline{m}_n = \begin{cases} n \text{ if } \underline{j}_1 = j \text{ and } \underline{j}_i = j \text{ or } \underline{j}_{i-1} = \underline{j}_{i+1} = j \\ \qquad\qquad \text{for all } i \text{ with } 1 < i < n, \\ n-1 \text{ if } \underline{j}_1 \neq j,\ \underline{j}_2 = j \text{ and } \underline{j}_i = j \text{ or } \underline{j}_{i-1} = \underline{j}_{i+1} = j \\ \qquad\qquad \text{for all } i \text{ with } 2 < i < n, \\ \min\{m \mid m \geq 1,\ \underline{j}_{n-m} \neq j,\ \underline{j}_{n-m-1} \neq j\} \text{ otherwise.} \end{cases}$$

We define $\underline{k}_n$ as the number of indices i for which $n-\underline{m}_n < i \leq n$ and $\underline{j}_i = j$. The n-th customer is called the *k-th of a j-sequence* if $\underline{k}_n = k$, $\underline{j}_n = j$.

We now have

$$\check{G}_n(\tau|j) = \sum_{m,k} \check{G}_n(\tau|j,m,k)P\{m,k|j\}$$

where the events $\underline{j}_n = j$, $\underline{m}_n = m$, $\underline{k}_n = k$ have been abbreviated to j, m, k, respectively, and where the sum is over all pairs (m,k) with $1 \leq k \leq m \leq n$.

Writing p instead of p_j and q instead of $1-p_j$, it is easily shown that

$$P\{m,k|j\} = \begin{cases} \binom{k-1}{n-k} p^{k-1} q^{n-k} & \text{if } m = n, \\ \binom{k-1}{n-k-1} p^{k-1} q^{n-k} & \text{if } m = n-1, \\ \binom{k-1}{m-k} p^{k-1} q^{m-k+2} & \text{if } m \leq n-2. \end{cases}$$

For example, if $m \leq n-2$, then the event $\underline{j}_n = j$, $\underline{m}_n = m$, $\underline{k}_n = k$ occurs if and only if $\underline{j}_n = j$, $\underline{j}_{n-m} \neq j$, $\underline{j}_{n-m-1} \neq j$, $\underline{j}_i = j$ for k-1 values of i with $n-m \leq i < n$ and there is at most one non-j-customer between each pair of otherwise successive j-customers. So there are k-1 spaces between these j-customers, each of which spaces may or may not be occupied by a non-j-customer, under the restriction that a total of m-k spaces is occupied, and the formula for P{m,k|j} if $m \leq n-2$ follows.

The other cases (m=n and m=n-1) can be proved similarly.

Hence, observing that $\check{G}_n(\tau|j,m,k)$ is independent of k, we have

$$\check{G}_n(\tau|j) = \check{G}_n(\tau|j,n) \sum_{k\geq 1} \binom{k-1}{n-k} p^{k-1} q^{n-k} + \\ + \check{G}_n(\tau|j,n-1) \sum_{k\geq 1} \binom{k-1}{n-k-1} p^{k-1} q^{n-k} + \\ + \sum_{m\leq n-2} \check{G}_n(\tau|j,m) \sum_{k\geq 1} \binom{k-1}{m-k} p^{k-1} q^{m-k+2}.$$

LEMMA 4.2. *Let* N *be a natural number, and* x *and* y *positive reals. Then* [1)]

$$\sum_{k\geq 1} \binom{k-1}{N-k} x^{k-1} y^{N-k} = \frac{1}{\sqrt{x^2+4xy}} \left\{ \left(\frac{x+\sqrt{x^2+4xy}}{2}\right)^N - \left(\frac{x-\sqrt{x^2+4xy}}{2}\right)^N \right\}.$$

PROOF. Define $f_N(x) = \sum_{k\geq 1} \binom{k-1}{N-k} x^k$. Then from

$$\binom{k-1}{N-k} = \binom{k-2}{N-k} + \binom{k-2}{N-k-1}$$

it follows that $f_N(x) = xf_{N-1}(x) + xf_{N-2}(x)$, a difference equation for f_N (with x as a parameter), which can be solved by standard methods, giving $f_N(x) = c_1\lambda_1^N + c_2\lambda_2^N$ with $\lambda_{1,2} = \frac{1}{2}(x \pm \sqrt{x^2+4x})$ and where c_1 and c_2 can be found from $f_1(x) = x$ and $f_2(x) = x^2$. To obtain the result of the lemma, it suffices to take xy^{-1} instead of x and to multiply by $x^{-1}y^N$. □

Applying the lemma and forming the generating function of $\check{G}_n(\tau|j)$, we obtain for $|u| < 1$:

$$\sum_{n\geq 2} \check{G}_n(\tau|j)u^n =$$

$$\frac{1}{\sqrt{p^2+4pq}} \left\{ \Sigma_1(w_1) - \Sigma_1(w_2) + q\Sigma_2(w_1) - q\Sigma_2(w_2) + q^2\Sigma_3(w_1) - q^2\Sigma_3(w_2) \right\}$$

where

$$\Sigma_1(w) = \sum_{n\geq 2} \check{G}_n(\tau|j,n) w^n u^n,$$

$$\Sigma_2(w) = \sum_{n\geq 2} \check{G}_n(\tau|j,n-1) w^{n-1} u^n,$$

$$\Sigma_3(w) = \sum_{n\geq 2} \sum_{m\leq n-2} \check{G}_n(\tau|j,m) w^m u^n,$$

$$w_1 = \frac{p+\sqrt{p^2+4pq}}{2}, \quad w_2 = \frac{p-\sqrt{p^2+4pq}}{2}.$$

Using the method of §3.21, $\Sigma_3(w)$ can be reduced as follows.

---

1) A binomial coefficient of the form $\binom{a}{b}$ with a and b non-negative integers has, by definition, the value 0 when b > a.

$$(4.17)\qquad \Sigma_3(w) = \sum_{n\geq 3}\sum_{m=1}^{n-2} \check{G}_n(\tau|j,m)w^m u^n =$$

$$= \sum_{m=1}^{\infty}\sum_{n=m+2}^{\infty} \check{G}_n(\tau|j,m)w^m u^n =$$

$$= \sum_{m=1}^{\infty}\sum_{h=2}^{\infty} \check{G}_{m+h}(\tau|j,m)(uw)^m u^h =$$

$$= \sum_{h=2}^{\infty} u^h \sum_{m=1}^{\infty} \check{G}_{m+h}(\tau|j,m)(uw)^m =$$

$$= \sum_{h=2}^{\infty} u^h \frac{\tau uw}{\lambda-\tau-\lambda uw}\left\{\frac{(\lambda-\tau)\check{G}_{h+1}(\tau|j;1)}{\tau} - [\tau := \lambda-\lambda uw]\right\} =$$

$$= \frac{\tau w}{\lambda-\tau-\lambda uw}\left\{\frac{\lambda-\tau}{\tau}\sum_{h=2}^{\infty} \check{G}_{h+1}(\tau|j;1)u^{h+1} - [\tau := \lambda-\lambda uw]\right\},$$

where we have evaluated the sum over m by applying lemma 3.1 with $\check{S} = 1$, $x = uw$.

Now $\check{G}_{h+1}(\tau|j;1) = \check{H}_{h+1}(\tau|\underline{j}_{h-1}\neq j, \underline{j}_h = j)$, and applying a small extension of lemma 4.1, we obtain

$$\check{G}_{h+1}(\tau|j;1) = \frac{\lambda\tau}{\lambda-\tau}\left\{\frac{\check{W}_h(\tau|\underline{j}_{h-1}\neq j)}{\tau} - [\tau := \lambda]\right\}.$$

Substituting this into (4.17), we obtain after some reduction

$$(4.18)\qquad \Sigma_3(w) = \frac{\lambda\tau uw}{\lambda-\tau-uw}\left\{\frac{1}{\tau}\sum_{h=2}^{\infty} \check{W}_h(\tau|\underline{j}_{h-1}\neq j)u^{h+1} - [\tau := \lambda-\lambda uw]\right\}.$$

If we denote $\lim_{u\to 1}(1-u)\Sigma_3(w)$ by $\Sigma_4(w)$, we have from (4.18):

$$\Sigma_4(w) = \frac{\lambda\tau w}{\lambda-\tau-w}\left\{\frac{1}{\tau}\check{W}(\tau|\underline{j}_{-1}\neq j) - [\tau := \lambda-\lambda w]\right\}$$

where $\check{W}(\tau|\underline{j}_{-1}\neq j)$ is given by lemma 3.3 in principle.

Since the analogous contributions of $\Sigma_1(w)$ and $\Sigma_2(w)$ are 0 in the limit, we have simply

$$\check{G}(\tau|j) = \frac{q^2}{\sqrt{p^2+4pq}}\left\{\Sigma_4(w_1)-\Sigma_4(w_2)\right\}.$$

The extension to $K \geq 2$ is not so simple here as it is in §4.32.

4.34 Concluding remarks

A related process of interest is what one might call "the full relaxation of the restricted process". By this we mean the process that arises when each customer fills as soon as possible with the proviso that in doing so he does not cause any customer to fill later than in the restricted process. The full relaxation is a very complicated process and we have obtained no results for it. Even in the special case of exponentially distributed emptying-times our attempts have been without success.

A process that looks much more promising is obtained by imposing "opportunist filling-priorities". For concreteness' sake we take $K = 2$, $M = 3$. The filling-discipline is then as follows. When a customer of type j arrives, he may fill as soon as he can (i.e. at once, provided the system does not contain the other two types), giving preference to a buffer containing j over an empty buffer. Emptying is in order of filling.

Suppose we are interested in the waiting-time of the customers. The resulting process can be analyzed by first neglecting the j-customers entirely. If $j = 1$, say, the time-axis is partitioned into intervals of 4 kinds: idle periods, 2-periods, 3-periods and 2-3-periods. (We speak of a 2-period when <u>only</u> type 2 is in the buffers, etc.). Now when a 1-customer arrives, his waiting-time is 0, unless he arrives during a 2-3-period, in which case he has to wait till the end of that period. At that epoch a 1-2-period or a 1-3-period starts (the length of which will depend, among other things, on the number of 1-customers then present), and the above-mentioned partition of the time-axis is replaced by a different one. Using the technique applied in §2.3 (which is given in [GÖBEL 1969] in somewhat more detail), it seems feasible to obtain results on the waiting-time in this way.

CHAPTER 5

ONE FINITE BUFFER

5.1 INTRODUCTION

In this chapter we consider a system with one finite buffer of capacity A. The number M of types is 1, except in §5.5 where we consider a relatively simple case with $M > 1$.

Filling and emptying takes place in order of arrival. Emptying takes place whenever the buffer contains a positive amount. A precise definition of the filling-discipline will be given in the various sections. We assume that emptying takes place at unit rate ($\omega_j = 1$), which is not a severe restriction. The filling-time per unit of load is α_j with $0 \le \alpha_j < 1$.

We assume that the n-th customer ($n \ge 1$) arrives at time $\underline{a}_n = \underline{y}_0 + \underline{y}_1 + \dots + \underline{y}_{n-1}$ where each $\underline{y}_i$ is exponential with parameter λ.

In §5.5, where $M > 1$, we use p_j and $\underline{j}_n$ in the same sense as in the previous chapters and we make the usual assumptions about these quantities.

Throughout this chapter we assume that the n-th customer carries an amount $\underline{s}_n$ where $S_j(s) = P\{\underline{s}_n \le s | \underline{j}_n = j\}$ does not depend on n. The variables $\underline{y}_0, \underline{y}_1, \underline{y}_2, \dots, \underline{s}_1, \underline{s}_2, \underline{s}_3, \dots$ are assumed to be independent.

In §5.2 we discuss some overflow-models, by which we mean models in which a customer who encounters a full buffer, leaves the system. In §5.3 to 5.5, when a customer notices that the buffer is full, he slows down his filling and fills at the emptying-rate. Here we speak of *retention* models.

5.2 OVERFLOW MODELS

We consider models with $\alpha_1 = 0$ only. Let us define the filling discipline as follows. Suppose at time $\underline{a}_n - 0$, the buffer contains an amount $\underline{t}_n$. If $\underline{s}_n + \underline{t}_n \le A$, the n-th customer fills the buffer on his arrival and departs. If $\underline{s}_n + \underline{t}_n > A$, he puts an amount $A - \underline{t}_n$ in the buffer, while the remainder of his load (an amount $\underline{s}_n + \underline{t}_n - A$) is taken elsewhere. Here we have, in fact, a well-known model in dam theory. The amount $\underline{s}_n + \underline{t}_n - A$ is usually called the "overflow".

The waiting-time and the filling-time of each customer are 0. The sequence $\underline{t}_n$ is more interesting. It is treated in detail in Chapter 5 of [COHEN 1969]. We repeat some of the simplest results. The process $\underline{t}_n$ satisfies

$$(5.1) \qquad \underline{t}_{n+1} = \max(0,\min(\underline{t}_n+\underline{s}_n,A)-\underline{y}_n).$$

Let the corresponding unbounded process $\underline{h}_n$ be defined by

$$(5.2) \qquad \begin{cases} \underline{h}_1 = 0, \\ \underline{h}_{n+1} = \max(0,\underline{h}_n+\underline{s}_n-\underline{y}_n). \end{cases}$$

If $\lambda\mathcal{E}\underline{s} < 1$, then the df of $\underline{t}_n$ tends to a limit $T(w)$ which is given by

$$(5.3) \qquad T(w) = \begin{cases} \dfrac{H(w)}{H(A)} & (w < A), \\ 1 & (w \geq A), \end{cases}$$

where H is the limiting df of $\underline{h}_n$ (given by the Pollaczek-Khintchine formula). Note that $T(w)$ is continuous for all $w > 0$, in particular at $w = A$.
Moran and Prabhu, among others, have considered a related overflow model, with a constant time (1, say) between successive arrivals. Effectively, their starting point is a process $\underline{t}_n$ defined by the relation

$$(5.4) \qquad \underline{t}_{n+1} = \max(0,\min(\underline{t}_n+\underline{s}_n,k)-m),$$

where k is the capacity of the buffer (or dam), m ($< k$) is the size of the desired release, and $\underline{t}_n$ the amount in the buffer at time n-0. The relations (5.1) and (5.4) are very similar. For results on this model we refer to [MORAN 1954] and [PRABHU 1958]. A simple relation for the limiting df of $\underline{t}_n$, analogous to (5.3), does not seem to exist in this model.

If we use the notation $\text{med}(\underline{a},\underline{b},\underline{c})$ for the median of $\underline{a},\underline{b},\underline{c}$, then (5.4) may be written as

$$(5.5) \qquad \underline{t}_{n+1} = \text{med}(0,\underline{t}_n+\underline{s}_n-m,k-m).$$

In this form, Moran's model has a close resemblance to an example given in [KEILSON 1963]. Keilson[1] considers the process

$$(5.6) \qquad \underline{t}_{n+1} = \text{med}(0,\underline{t}_n+\underline{u}_n,A),$$

---

[1] See also [COHEN, 1969], p.466-494.

where $\underline{u}_n$ has a density $\phi(u)$ on R^1 which is for negative values of u of the form $c_1 e^{c_2 u}$ and where $\mathcal{E}\underline{u}_n < 0$. It is easy to see that the choice $\underline{u}_n = \underline{s}_n - \underline{y}_n$ where $\underline{s}_n$ and $\underline{y}_n$ are independent and $\underline{y}_n$ is exponentially distributed with $\mathcal{E}\underline{y}_n < \mathcal{E}\underline{s}_n$ satisfies this condition. He obtains the following result.

$$(5.7) \qquad T(w) = \begin{cases} c\,\dfrac{H(w)}{H(A)} & (w < A), \\ 1 & (w \geq A), \end{cases}$$

where c is a constant (which depends on A) with $H(A) < c < 1$ and where H is, as before, the limiting df of the unrestricted process. In this case, T has a jump at A.

In the special case where both $\underline{s}_n$ and $\underline{y}_n$ have exponential distibutions, one obtains the following explicit result. Let $\mathcal{E}\underline{s}_n = \mu^{-1}$, $\mathcal{E}\underline{y}_n = \lambda^{-1}$, $\rho = \lambda\mu^{-1}$. Then for $0 \leq w < A$, $T(w)$ is given by

$$(5.8) \qquad T(w) = \frac{1 - \rho e^{-(\mu-\lambda)w}}{1 - \rho^2 e^{-(\mu-\lambda)A}} \qquad \text{if } \lambda \neq \mu,$$

and by

$$(5.9) \qquad T(w) = \frac{1 + \lambda w}{2 + \lambda A} \qquad \text{if } \lambda = \mu.$$

Note that $\rho < 1$ is not required here. When $\rho = 1$, the limiting distribution is "H-shaped" (a mass $\frac{1}{2+\lambda A}$ in 0 and in A, and a uniform distribution in between).

5.3 A RETENTION-MODEL WITH INFINITE FILLING-RATE

In this section we again take $M = 1$, $\alpha_1 = 0$. We define the filling-discipline as follows. The n-th customer waits until the filling-line is no longer used by customer n-1, and then (at time $\underline{b}_n$) he fills or starts to fill. Let the amount in the buffer at time $\underline{b}_n - 0$ be $\underline{t}_n$. If $\underline{t}_n + \underline{s}_n \leq A$, the n-th customer fills the buffer instanteously and departs. If $\underline{t}_n < A$ and $\underline{t}_n + \underline{s}_n > A$, the n-th customer puts an amount $A - \underline{t}_n$ in the buffer at once, and the remainder of his load at a rate 1 (viz. the emptying-rate). If $\underline{t}_n = A$, the n-th

customer "fills" the buffer at a rate 1.
In the subsections, we consider the waiting time $\underline{w}_n$, the process $\underline{t}_n$, and the filling-time $\underline{r}_n$.

5.31 The waiting-time

Let $\underline{w}_n = \underline{b}_n - \underline{a}_n$ be the waiting-time of the n-th customer and $\underline{h}_n = \underline{c}_n - \underline{a}_n$ the waiting-time of the n-th load, where $\underline{c}_n$ is the epoch at which the emptying-line starts operating on the n-th load.
We further introduce the epochs $\underline{d}_n$ and $\underline{e}_n$, the end of the filling and the emptying-operation, respectively.
We then have the following simple relations

$$\underline{b}_{n+1} = \max(\underline{a}_{n+1}, \underline{d}_n),$$

$$\underline{c}_{n+1} = \max(\underline{a}_{n+1}, \underline{e}_n),$$

$$\underline{d}_n = \underline{b}_n + \underline{r}_n,$$

$$\underline{e}_n = \underline{c}_n + \underline{s}_n.$$

Also, $\underline{h}_{n+1} = \underline{c}_{n+1} - \underline{a}_{n+1} = \max(0, \underline{e}_n - \underline{a}_n - \underline{y}_n) = \max(0, \underline{c}_n + \underline{s}_n - \underline{a}_n - \underline{y}_n)$, hence (cf. (5.2))

$$\underline{h}_{n+1} = \max(0, \underline{h}_n + \underline{s}_n - \underline{y}_n). \tag{5.10}$$

It follows that the process $\underline{h}_n$ does not depend on A. This is intuitively clear.
In the same way, one may prove that

$$\underline{w}_{n+1} = \max(0, \underline{w}_n + \underline{r}_n - \underline{y}_n), \tag{5.10a}$$

but this relation is virtually useless since the filling-times $\underline{r}_1, \underline{r}_2, \ldots$ are mutually dependent (cf. §5.33). However, $\underline{w}_n$ can be obtained at once from $\underline{h}_n$:

$$\underline{w}_n = \max(0, \underline{h}_n - A). \tag{5.11}$$

This can be shown as follows. Since $\omega = 1$, $\underline{h}_n$ can be interpreted as the amount present in the *system* (i.e. in the buffer or carried by a customer) at the epoch $\underline{a}_n - 0$. If $\underline{h}_n < A$, filling can start at once, hence $\underline{w}_n = 0$. If $\underline{h}_n \geq A$, the n-th customer has to wait until an amount $\underline{h}_n - A$ has been removed from the system, and (5.11) follows.

Now suppose $\lambda \mathcal{E}\underline{s} < 1$. Then $\underline{h}_n$ has a limiting distribution as $n \to \infty$ with LST given by the Pollaczek-Khintchine formula

$$\check{H}(\tau) = \frac{(1-\rho)\tau}{\tau-\lambda+\lambda\check{S}(\tau)}$$

where $\rho = \lambda\mathcal{E}\underline{s} = \lambda/\mu$.
Consequently, $\underline{w}_n$ has a limiting distribution; we denote its LST by $\check{W}(\tau,A)$. We introduce $\underline{h}$ as a random variable with LST $\check{H}(\tau)$, which is independent of $\underline{s}_1,\underline{s}_2,\ldots,\underline{y}_0,\underline{y}_1,\ldots$. Then we can write

$$(5.12) \qquad \check{W}(\tau,A) = \mathcal{E}e^{-\tau\max(0,\underline{h}-A)} = \mathcal{E}e^{-\tau\underline{w}},$$

where $\underline{w} = \max(0,\underline{h}-A)$.
When S is of a simple form, $\check{W}(\tau,A)$ or even the df of $\underline{w}$, can be determined explicitely. In other cases, we may rewrite (5.12) in a different form which can be more useful. We proceed as follows.
We restrict τ to non-negative reals. As a function of A, $\check{W}(\tau,A)$ is defined for all real A, it is non-decreasing and continuous; some values are $\check{W}(\tau,-\infty) = 0$ (provided $\tau > 0$), $\check{W}(\tau,0) = \check{H}(\tau)$, $\check{W}(\tau,\infty) = 1$.

LEMMA 5.1. *The derivative of* $\check{W}(\tau,A)$ *with respect to* A *exists for all real* $A \neq 0$, *and is given by*

$$\tau e^{\tau A}\int_{\max(A,0)}^{\infty} e^{-\tau h}dH(h).$$

PROOF. If $A < 0$, it follows from (5.12) that $\check{W}(\tau,A) = e^{\tau A}\check{H}(\tau)$, hence the statements of the lemma.
Hence suppose that $A > 0$. Let dA be a (small) positive number. Then

$$\check{W}(\tau,A+dA) - \check{W}(\tau,A) = \int_{0^-}^{\infty}\left\{e^{-\tau\max(0,h-A-dA)}-e^{-\tau\max(0,h-A)}\right\}dH(h).$$

When $h < A$, the integrand is 0. When $A \leq h < A+dA$, the contribution to the integral is o(dA), hence negligible. (This rests on the continuity of H for $h > 0$.) When $h \geq A+dA$, the integrand may be written as

$$e^{-\tau(h-A)}\{1+\tau dA+O(dA)^2-1\},$$

and we have

$$\check{W}(\tau,A+dA) - \check{W}(\tau,A) = \int_{A+dA}^{\infty} e^{-\tau(h-A)}\{\tau dA+O(dA)^2\}dH(h),$$

hence

$$\lim_{dA\downarrow 0} \frac{\check{W}(\tau,A+dA)-\check{W}(\tau,A)}{dA} = \tau\int_{A}^{\infty} e^{-\tau(h-A)}dH(h).$$

A similar argument can be used to obtain the left hand derivative, and the statements of the lemma for $A > 0$ follow. □

We now define a transform of $\check{W}(\tau,A)$ with respect to A by

$$(5.13)\qquad \check{\check{W}}(\tau,\sigma) \overset{\text{def}}{=} \int_{0}^{\infty} e^{-\sigma A}d_A\check{W}(\tau,A), \qquad (\sigma \geq 0).$$

Note that $\check{\check{W}}(\tau,0)$ is not 1, but $1-\check{H}(\tau)$.
From (5.13) and lemma 5.1 it follows that

$$\check{\check{W}}(\tau,\sigma) = \int_{0^+}^{\infty} e^{-\sigma A}\tau e^{\tau A}\int_{A}^{\infty} e^{-\tau h}dH(h)dA =$$
$$= \tau\int_{h=0}^{\infty}\left(\int_{A=0}^{h} e^{(\tau-\sigma)A}dA\right)e^{-\tau h}dH(h),$$

which reduces to

$$(5.14)\qquad \check{\check{W}}(\tau,\sigma) = -\tau\,\frac{\check{H}(\tau)-\check{H}(\sigma)}{\tau-\sigma}\,.$$

This formula can be used for example to obtain a transform of $\mathcal{E}\underline{w}$, by differentiating (5.14) with respect to τ and letting $\tau \to 0$. The result is

$$(5.15)\qquad \int_{0}^{\infty} e^{-\sigma A}d_A\mathcal{E}\underline{w} = -\frac{1-\check{H}(\sigma)}{\sigma}\,.$$

The various interchanges of operations, required to obtain this result, are allowed since the functions involved are non-negative.

The *transient behaviour* of $\underline{w}_n$ may be treated as follows. In analogy to (5.14) we have

$$(5.16) \qquad \check{\check{W}}_n(\tau,\sigma) = -\tau \frac{\check{H}_n(\tau)-\check{H}_n(\sigma)}{\tau-\sigma}$$

and hence, for $|x| < 1$:

$$W(\tau,\sigma,x) \overset{\text{def}}{=} \sum_{n=1}^{\infty} \check{\check{W}}_n(\tau,\sigma)x^n = -\frac{\tau}{\tau-\sigma}\left\{\sum_{n=1}^{\infty} \check{H}_n(\tau)x^n - [\tau := \sigma]\right\}.$$

Lemma 3.1 can now be applied to $\sum_{n=1}^{\infty}\check{H}_n(\tau)x^n$, and we obtain

$$(5.17) \qquad W(\tau,\sigma,x) = \frac{-\tau x}{\tau-\sigma}\left\{\frac{\tau}{\lambda-\tau-\lambda x\check{S}(\tau)}\left\{\frac{\lambda-\tau}{\tau}\check{H}_1(\tau) - [\tau := \lambda-\lambda z]\right\} - [\tau := \sigma]\right\},$$

where z is usual root of $z = x\check{S}(\lambda-\lambda z)$.

5.32 The amount in the buffer

In §5.3, $\underline{t}_n$ has been defined as the amount in the buffer at $\underline{b}_n$-0, i.e. just before the filling starts. An equivalent definition would have been "the amount in the buffer at $\underline{a}_n$-0". For, if $\underline{a}_n = \underline{b}_n$, this is trivially true. But $\underline{a}_n < \underline{b}_n$ means that the n-th customer has to wait before he can start filling, which implies that the buffer is full at the epoch $\underline{a}_n$-0; hence at $\underline{b}_n$-0. So if $\underline{a}_n \neq \underline{b}_n$, then $\underline{t}_n = A$ according to both definitions, hence in this case, too, the definitions are equivalent. This observation implies the simple relation

$$(5.18) \qquad \underline{t}_n = \min(\underline{h}_n, A),$$

and if $\lambda\mathcal{E}\underline{s} < 1$, $\underline{t}_n$ has a limiting df T given by

$$(5.19) \qquad T(w) = \begin{cases} H(w) & (w < A), \\ 1 & (w \geq A). \end{cases}$$

A close relation exists between (5.3), (5.7) and (5.19). E.g. from (5.19), it follows that

$$P\{\underline{t} \le w | \underline{t} < A\} = \begin{cases} \dfrac{H(w)}{H(A)} & (w < A), \\ 1 & (w \ge A), \end{cases}$$

and the right hand side coincides with that of (5.3). This is intuitively clear if we take into account the following facts:

1) whenever $\underline{t}_n+\underline{s}_n < A$, the value of $\underline{t}_{n+1}$ as defined by (5.1) coincides with the value of $\underline{t}_{n+1}$ as defined by (5.10) and $\underline{t}_{n+1} = \min(\underline{h}_{n+1},A)$;
2) the present process can be given the property $P\{\underline{t}_n = A\} = 0$ by removing from the time axis the intervals during which $\underline{t}_n = A$;
3) the variables $\underline{y}_n$ are exponential.

In the same way, one may guess (5.3) from (5.7), or (5.7) from (5.19).

From (5.11) and (5.18) it follows that

(5.20) $$\underline{t}_n = \underline{h}_n - \underline{w}_n.$$

Since $\underline{t}_n$ and $\underline{w}_n$ are so closely related, it does not seem desirable to treat $\underline{t}_n$ extensively. We just give the following results.
If

(5.21) $$\check{T}(\tau,A) \overset{\text{def}}{=} \mathcal{E}e^{-\tau\underline{t}}$$

where $\underline{t}$ is a random variable with df T, and if

(5.22) $$\check{\check{T}}(\tau,\sigma) \overset{\text{def}}{=} \int_0^\infty e^{-\sigma A} d_A\check{T}(\tau,A),$$

then

(5.23) $$\check{\check{T}}(\tau,\sigma) = \frac{-\tau}{\tau+\sigma}\{1 - \check{H}(\sigma+\tau)\}$$

and

(5.24) $$\int_0^\infty e^{-\sigma A} d_A \mathcal{E}\underline{t} = \frac{1 - \check{T}(\sigma)}{\sigma}.$$

5.33 The filling-time

It is easily seen that $\underline{r}_n$ satisfies the relations

(5.25) $$\underline{r}_n = \max(0,\underline{s}_n+\underline{t}_n-A),$$
(5.26) $$\underline{r}_n = \text{med}(0,\underline{s}_n+\underline{h}_n-A,\underline{s}_n).$$

From (5.26), one can obtain the transform

$$\check{\check{R}}(\tau,\sigma) = \int_0^\infty e^{-\sigma A} d_A \mathcal{E} e^{-\tau \underline{r}}$$

where $\underline{r}$ has an obvious meaning. Along the lines of the proof of (5.14), we find

$$\check{\check{R}}(\tau,\sigma) = \frac{\tau}{\tau-\sigma} \check{H}(\sigma)\{\check{S}(\sigma)-\check{S}(\tau)\}. \qquad (5.27)$$

5.4 A RETENTION MODEL WITH FINITE FILLING-RATE

In this section we take, as before, $M = 1$ but we assume now that the filling-rate is finite. In fact, dropping the index 1, we assume $0 < \alpha < 1$.
The filling-discipline is defined as follows. The n-th customer waits until the filling-line is no longer in use by customer n-1, and starts to fill (at time $\underline{b}_n$). He fills at a rate α^{-1} as long as the buffer is not full; when the buffer is or becomes full, he fills at a rate 1.
It is interesting to compare the basic relations in §5.3 with those in the present section.
The waiting-time $\underline{h}_n$ till the start of the emptying-operation still satisfies

$$\underline{h}_{n+1} = \max(0,\underline{h}_n+\underline{s}_n-\underline{y}_n). \qquad (5.28)$$

Likewise, in analogy to (5.10a):

$$\underline{w}_{n+1} = \max(0,\underline{w}_n+\underline{r}_n-\underline{y}_n). \qquad (5.29)$$

But instead of (5.11), we now have

$$\underline{w}_{n+1} = \max(0,\underline{w}_n+\alpha\underline{s}_n-\underline{y}_n,\underline{h}_{n+1}-A). \qquad (5.30)$$

This will be proved presently. First we need the relations for $\underline{r}_n$ and $\underline{t}_n$.
It is easily seen that $\underline{r}_n$ satisfies

$$\underline{r}_n = \max(\alpha\underline{s}_n,\underline{s}_n+\underline{t}_n-A). \qquad (5.31)$$

We further claim that

(5.32) $\quad \underline{t}_n = \underline{h}_n - \underline{w}_n.$

This can be shown as follows. Let $\underline{b}_n$ and $\underline{c}_n$ be defined as in §5.3. Then $\underline{h}_n-\underline{w}_n = \underline{c}_n-\underline{b}_n$. Now suppose $\underline{c}_n-\underline{b}_n > A$. During the interval $[\underline{b}_n,\underline{c}_n)$ the emptying-line is busy only on loads numbered n-1 or lower. These loads have been put in the system prior to $\underline{b}_n$. This leads to a contradiction and we conclude that $\underline{c}_n-\underline{b}_n \le A$. Formula (5.32) now follows at once from the obvious fact $\underline{t}_n = \min(\underline{h}_n-\underline{w}_n,A)$.

To prove (5.30), we substitute (5.31) and (5.32) into (5.29):

$$\begin{aligned}\underline{w}_{n+1} &= \max(0,\underline{w}_n+\alpha\underline{s}_n-\underline{y}_n,\underline{w}_n+\underline{s}_n+\underline{t}_n-A-\underline{y}_n) = \\ &= \max(0,\underline{w}_n+\alpha\underline{s}_n-\underline{y}_n,\underline{s}_n+\underline{h}_n-A-\underline{y}_n) = \\ &= \max(0,\underline{w}_n+\alpha\underline{s}_n-\underline{y}_n,\underline{h}_{n+1}-A).\end{aligned}$$

The last step is not quite obvious. It amounts to saying that if A, X, Y are reals with $A \ge 0$, then

(*) $\quad \max(0,X,Y-A) = \max(0,X,\max(Y,0)-A).$

In fact, if $Y \ge 0$, this is immediate, while if $Y < 0$, then $Y-A < 0$ and $\max(Y,0)-A \le 0$, so that the maximum in (*) is 0 or X, both on the left and right hand side.
This completes the proof of (5.30).

Let $\underline{u}_i = \underline{s}_i-\underline{y}_i$, $\underline{v}_i = \alpha\underline{s}_i-\underline{y}_i$, and suppose that $\underline{h}_1$ and $\underline{w}_1$ are identically 0. Then we can prove, in the same way as in Chapter 2 that the pair $(\underline{h}_n,\underline{w}_n)$ has the same df as the pair $(\underline{h}'_n,\underline{w}'_n)$ where

(5.33) $\quad \underline{h}'_n = \max_{0\le k\le n-1} \{\underline{u}_1+\ldots+\underline{u}_{k-1}\}$

and

(5.34) $\quad \underline{w}'_n = \max_{0\le i\le j\le n-1} \{\underline{v}_1+\ldots+\underline{v}_i+\underline{u}_{i+1}+\ldots+\underline{u}_j-A\kappa_{ij}\}$

where κ_{ij} is 1 minus Kronecker's δ_{ij}.
The same arguments as in Chapter 2 now show that the limits

$$W(w) = \lim_{n\to\infty} P\{\underline{w}_n \le w\}$$

and

$$F(h,w) = \lim_{n\to\infty} P\{\underline{h}_n \le h;\ \underline{w}_n \le w\}$$

exist, and that these limits are distribution functions provided $\mathcal{E}\underline{u}_1 < 0$ and $\mathcal{E}\underline{v}_1 < 0$. Furthermore it can be shown that F satisfies the integral equation

$$(5.35) \qquad F(h,w) = \iint F(h\wedge(A+w)+y-s, w+y-\alpha s)\,dS(s)\lambda e^{-\lambda y}dy.$$

5.5 A RETENTION MODEL WITH SEVERAL TYPES OF CUSTOMERS

In this section we take $M > 1$ and $\alpha_1=\ldots=\alpha_M=0$. The filling-discipline is defined as follows. If the n-th customer is of the same type as his predecessor, he behaves as in §5.3. If the n-th customer is not of the type of his predecessor, he waits until the buffer is empty, then puts an amount $\min(\underline{s}_n,A)$ in the buffer instantaneously, and then a possible rest of his load, of size $\max(0,\underline{s}_n-A)$, at a rate 1.

The *waiting-time* of the n-th customer until filling is called $\underline{w}_n$. The conditional LST of $\underline{w}_n$ under the condition $\underline{j}_n = j$ is denoted by $\check{W}_n(\tau,A|j)$. We define $\underline{k}_n$ in the same way as in §3.21. Then

$$\check{W}_n(\tau,A|j) = \sum_{k=1}^{n} \mathcal{E}\left(e^{-\tau\underline{w}_n}\middle|j,k\right)P\{\underline{k}_n=k|j\} =$$
$$= \sum_{k=1}^{n-1} \mathcal{E}\left(e^{-\tau\underline{w}_n}\middle|j,k\right)q_jp_j^{k-1}+\mathcal{E}\left(e^{-\tau\underline{w}_n}\middle|j,n\right)p_n^{n-1}.$$

Forming the generating function we obtain

$$W(\tau,A,x|j) \overset{\text{def}}{=} \sum_{n=1}^{\infty} W_n(\tau,A|j)x^n = \Sigma_1 + \Sigma_2,$$

where

$$\Sigma_1 = \sum_{n=2}^{\infty}\sum_{k=1}^{n-1} \mathcal{E}\left(e^{-\tau\underline{w}_n}\middle|j,k\right)q_jp_j^{k-1}x^n,$$

$$\Sigma_2 = \sum_{n=1}^{\infty} \mathcal{E}\left(e^{-\tau\underline{w}_n}\middle|j,n\right)p_j^{n-1}x^n.$$

In Σ_1, we may change the order of summation, substitute $n = k+m$, and change the order of summation again, obtaining

$$\Sigma_1 = \sum_{m=1}^{\infty} q_j p_j^{-1} x^m \sum_{k=1}^{\infty} \mathscr{E}\left(e^{-\tau \underline{w}_{k+m}} \middle| j,k\right)(p_j x)^k.$$

Transforming with respect to A, we obtain

$$(5.36) \qquad \int_0^{\infty} e^{-\sigma A} d_A \Sigma_1 = \sum_{m=1}^{\infty} q_j p_j^{-1} x^m \sum_{k=1}^{\infty} \int_0^{\infty} e^{-\sigma A} d_A \left\{\mathscr{E}\left(e^{-\tau \underline{w}_{k+m}} \middle| j,k\right)\right\}(p_j x)^k.$$

If $k = 1$, the integral in (5.36) is 0, since $\mathscr{E}(e^{-\tau \underline{w}_{m+1}}|j,1)$ is equal to $\mathscr{E}(e^{-\tau \underline{h}_{m+1}}|j,1)$, where $\underline{h}_{m+1}$ is as in §5.3, and the df of $\underline{h}_{m+1}$ does not depend on A.

If $k > 1$, the (k+m)-th customer is *not* the first of a j-sequence, and hence, applying a slight generalization of (5.16), the integral in (5.36) can be written as

$$\check{W}_{k+m}(\tau,\sigma|j,k) = -\frac{\tau}{\tau-\sigma}\{\check{H}_{k+m}(\tau|j,k)-[\tau := \sigma]\}.$$

Substituting this into (5.36), we obtain

$$\int_0^{\infty} e^{-\sigma A} d_A \Sigma_1 = -\frac{q_j}{p_j}\cdot\frac{\tau}{\tau-\sigma}\sum_{m=1}^{\infty} x^m \left\{\sum_{k=2}^{\infty} \check{H}_{k+m}(\tau|j,k)(p_j x)^k-[\tau := \sigma]\right\}.$$

The sum over k can now be reduced by applying lemma 3.1, giving

$$\int_0^{\infty} e^{-\sigma A} d_A \Sigma_1 = -\frac{q_j}{p_j}\cdot\frac{\tau}{\tau-\sigma}\cdot\sum_{m=1}^{\infty} x^m \cdot$$

$$(5.37) \qquad \cdot\left\{\frac{\tau p_j x}{\lambda-\tau-\lambda p_j x \check{S}_j(\tau)}\left\{\frac{\lambda-\tau}{\tau}\check{H}_{m+1}(\tau|j)-[\tau := \lambda-\lambda z_j]\right\} + \right.$$

$$\left. -\check{H}_{m+1}(\tau|j)p_j x-[\tau := \sigma]\right\}$$

where z_j is the root with smallest absolute value of

$$z_j = p_j x \check{S}_j(\lambda-\lambda z_j).$$

The sum Σ_2 can be treated in a similar manner, but the result has no influence upon the limit we are interested in, viz.

$$\check{W}(\tau,A|j) \overset{\text{def}}{=} \lim_{n\to\infty} \check{W}_n(\tau,A|j).$$

The existence of this limit can be shown using exactly the same argument as in §3.21. Hence the transform

$$\check{\check{W}}(\tau,\sigma|j) \overset{\text{def}}{=} \int_0^\infty e^{-\sigma A} d_A \check{W}(\tau,A|j)$$

exists, too.

Since $\lim\limits_{x\uparrow 1}(1-x)\Sigma_2 = 0$, as indicated earlier, we can now obtain $\check{\check{W}}(\tau,\sigma|j)$ from (5.37) by applying Abel's theorem performing a permitted change of operators:

$$\check{\check{W}}(\tau,\sigma|j) = \int_0^\infty e^{-\sigma A} d_A\left(\lim_{x\uparrow 1}(1-x)\Sigma_1\right) = \lim_{x\uparrow 1}(1-x)\int_0^\infty e^{-\sigma A} d_A \Sigma_1 .$$

From (5.36) we then obtain the final result:

$$\check{\check{W}}(\tau,\sigma|j) = \frac{-q_j\tau}{\tau-\sigma}\left\{\frac{\tau}{\lambda-\tau-\lambda p_j S_j(\tau)}\left\{\frac{\lambda-\tau}{\tau}\check{H}(\tau|j)-[\tau := \lambda-\lambda z_j]\right\}-\check{H}(\tau|j)-[\tau := \sigma]\right\}$$

where z_j is the root with smallest absolute value of

$$z_j = p_j \check{S}_j(\lambda-\lambda z_j).$$

SUMMARY

A great variety of queueing problems exist that can be adequately described by a mathematical model involving one or more buffers. An example: the customers are oil-tankers arriving at a refinery where the various types of crude oil have to be stored temporarily. The practical problem here is to choose the number and sizes of the buffers such that the cost of the buffers and the waits of the tankers are balanced.

In order that a model of such a complex situation be manageable, certain simplifying assumptions have to be made. E.g., in the greater part of our treatise we have assumed that the buffers have infinite capacity. Even then, questions on the waiting-time of the customers, say, can be answered only in special cases.

We have divided the various possibilities into four groups, corresponding to the chapters 2,3,4,5. In chapter 2, there are as many infinite buffers as there are types of customers. This assumption entails trivial answers to certain obvious questions, thereby inviting other questions. As a result, chapter 2 stands a little apart from the remainder.

In chapter 3, or more precisely in §3.21, the central result is derived, giving the waiting-time of a customer in a model where an arbitrary number of types of customers share one infinite buffer. The rest of chapter 3 presents some variations on this theme.

In chapters 4 and 5, where the results of §3.21 are applied repeatedly, we consider some models involving several infinite buffers and one finite buffer, respectively.

REFERENCES

AVI-ITZHAK, B., MAXWELL, W.L., and L.W. MILLER (1965), *Queueing with alternating priorities.* J. Oper. Res. Soc. Am. 13 306-318.

AVI-ITZAK, B., and P. NAOR (1963), *Some queueing problems with the service station subject to breakdown.* J. Oper. Res. Soc. Am. 11 303-320.

COBHAM, A. (1954), *Priority assignment in waiting-line problems.* J. Oper. Res. Soc. Am. 2 70-76.

COHEN, J.W. (1969), *The Single Server Queue.* North-Holland Publishing Company; Amsterdam.

COHEN, J.W. (1974), *Superimposed renewal processes and storage with gradual input.* Stoch. Proc. Appl. 2 31-57.

FELLER, W. (1966), *An Introduction to Probability Theory and Its Applications,* volume 2. John Wiley & Sons; New York.

FELLER, W. (1967), *An Introduction to Probability Theory and Its Applications,* volume 1, 3rd edition. John Wiley & Sons; New York.

GAVER, D.P. (1963), *A comparison of queue disciplines when service orientation times occur.* Naval Res. Logistics Qly. 10 219-235.

GÖBEL, F. (1965), *Some queueing models with dependent service times.* Math. Centrum, report S350 (VP25), Amsterdam.

GÖBEL, F. (1969), *A queueing model with opportunist priorities.* Math. Comm. Twente Univ. Techn. 4 (4).

GÖBEL, F. (1974), *Queueing models involving buffers.* Thesis. Mathematisch Centrum, Amsterdam.

JAISWAL, N.K. (1968), *Priority Queues.* Academic Press; New York.

JEWELL, W.S. (1967), *A simple proof of:* $L = \lambda W$. J. Oper. Res. Soc. Am. 15 1109-1116.

KEILSON, J. (1963), *Green's function methods in probability theory.* (Summary of three lectures compiled by W. Molenaar), Math. Centrum, report S318, Amsterdam.

KENDALL, D.G. (1951), *Some problems in the theory of queues.* J. Royal Stat. Soc. B13 151-185.

KESTEN, H., and J.Th. RUNNENBURG (1957), *Priority in waiting line problems.* Indagationes Math. 19 312-336.

KINGMAN, J.F.C. (1962), *The effect of queue discipline on waiting time variance.* Proc. Cambr. Phil. Soc. 58 163-164.

LINDLEY, D.V (1952), *The theory of queues with a single server.* Proc. Cambr. Phil. Soc. 48 277-289.

LITTLE, J.D.C. (1961), *A proof for the queueing formula* L = λW. J. Oper. Res. Soc. Am. 9 383-387.

MORAN, P.A.P. (1954), *A probability theory of dams and storage systems.* Austr. J. Appl. Sci. 5 116-124.

PRABHU, N.U. (1958), *On the integral equation for the finite dam.* Quart. J. Math. Oxford 9 183-188.

RUNNENBURG, J.Th. (1960), *On the Use of Markov Processes in One-Server Waiting-Time Problems and Renewal Theory.* Thesis & Propositions, Poortpers N.V., Amsterdam.

RUNNENBURG, J.Th. (1965), *On the use of the method of collective marks in queueing theory.* Ch. 13 in: Proc. Symp. Congestion Theory, W.L. Smith & W.E. Wilkinson, eds. Chapel Hill (1965).

SMITH, W.L. (1958), *Renewal theory and its ramifications.* J. Royal. Stat. Soc. B20 243-302.

TACKÁCS, L. (1962), *Introduction to the Theory of Queues.* Oxford Univ. Press; New York.

TACKÁCS, L. (1968), *Two queues attended by a single server.* J. Oper. Res. Soc. Am. 16 639-650.

INDEX

OTHER TITLES IN THE SERIES MATHEMATICAL CENTRE TRACTS

A leaflet containing an order-form and abstracts of all publications mentioned below is available at the Mathematisch Centrum, Tweede Boerhaavestraat 49, Amsterdam-1005, The Netherlands. Orders should be sent to the same address.

MCT 1 T. VAN DER WALT, *Fixed and almost fixed points*, 1963. ISBN 90 6196 002 9.

MCT 2 A.R. BLOEMENA, *Sampling from a graph*, 1964. ISBN 90 6196 003 7.

MCT 3 G. DE LEVE, *Generalized Markovian decision processes, part I: Model and method*, 1964. ISBN 90 6196 004 5.

MCT 4 G. DE LEVE, *Generalized Markovian decision processes, part II: Probabilistic background*, 1964. ISBN 90 6196 006 1.

MCT 5 G. DE LEVE, H.C. TIJMS & P.J. WEEDA, *Generalized Markovian decision processes, Applications*, 1970. ISBN 90 6196 051 7.

MCT 6 M.A. MAURICE, *Compact ordered spaces*, 1964. ISBN 90 6196 006 1.

MCT 7 W.R. VAN ZWET, *Convex transformations of random variables*, 1964. ISBN 90 6196 007 X.

MCT 8 J.A. ZONNEVELD, *Automatic numerical integration*, 1964. ISBN 90 6196 008 8.

MCT 9 P.C. BAAYEN, *Universal morphisms*, 1964. ISBN 90 6196 009 6.

MCT 10 E.M. DE JAGER, *Applications of distributions in mathematical physics*, 1964. ISBN 90 6196 010 X.

MCT 11 A.B. PAALMAN-DE MIRANDA, *Topological semigroups*, 1964. ISBN 90 6196 011 8.

MCT 12 J.A.TH.M. VAN BERCKEL, H. BRANDT CORSTIUS, R.J. MOKKEN & A. VAN WIJNGAARDEN, *Formal properties of newspaper Dutch*, 1965. ISBN 90 6196 013 4.

MCT 13 H.A. LAUWERIER, *Asymptotic expansions*, 1966, out of print; replaced by MCT 54.

MCT 14 H.A. LAUWERIER, *Calculus of variations in mathematical physics*, 1966. ISBN 90 6196 020 7.

MCT 15 R. DOORNBOS, *Slippage tests*, 1966. ISBN 90 6196 021 5.

MCT 16 J.W. DE BAKKER, *Formal definition of programming languages with an application to the definition of ALGOL 60*, 1967. ISBN 90 6196 022 3.

MCT 17 R.P. VAN DE RIET, *Formula manipulation in ALGOL 60, part 1*, 1968. ISBN 90 6196 025 8.

MCT 18 R.P. VAN DE RIET, *Formula manipulation in ALGOL 60, part 2*, 1968. ISBN 90 6196 038 X.

MCT 19 J. VAN DER SLOT, *Some properties related to compactness*, 1968. ISBN 90 6196 026 6.

MCT 20 P.J. VAN DER HOUWEN, *Finite difference methods for solving partial differential equations*, 1968. ISBN 90 6196 027 4.

MCT 21 E. WATTEL, *The compactness operator in set theory and topology*, 1968. ISBN 90 6196 028 2.

MCT 22 T.J. DEKKER, *ALGOL 60 procedures in numerical algebra, part 1*, 1968. ISBN 90 6196 029 0.

MCT 23 T.J. DEKKER & W. HOFFMANN, *ALGOL 60 procedures in numerical algebra, part 2*, 1968. ISBN 90 6196 030 4.

MCT 24 J.W. DE BAKKER, *Recursive procedures*, 1971. ISBN 90 6196 060 6.

MCT 25 E.R. PAERL, *Representations of the Lorentz group and projective geometry*, 1969. ISBN 90 6196 039 8.

MCT 26 EUROPEAN MEETING 1968, *Selected statistical papers, part I*, 1968. ISBN 90 6196 031 2.

MCT 27 EUROPEAN MEETING 1968, *Selected statistical papers, part II*, 1969. ISBN 90 6196 040 1.

MCT 28 J. OOSTERHOFF, *Combination of one-sided statistical tests*, 1969. ISBN 90 6196 041 X.

MCT 29 J. VERHOEFF, *Error detecting decimal codes*, 1969. ISBN 90 6196 042 8.

MCT 30 H. BRANDT CORSTIUS, *Excercises in computational linguistics*, 1970. ISBN 90 6196 052 5.

MCT 31 W. MOLENAAR, *Approximations to the Poisson, binomial and hypergeometric distribution functions*, 1970. ISBN 90 6196 053 3.

MCT 32 L. DE HAAN, *On regular variation and its application to the weak convergence of sample extremes*, 1970. ISBN 90 6196 054 1.

MCT 33 F.W. STEUTEL, *Preservation of infinite divisibility under mixing and related topics*, 1970. ISBN 90 6196 061 4.

MCT 34 I. JUHÁSZ, A. VERBEEK & N.S. KROONENBERG, *Cardinal functions in topology*, 1971. ISBN 90 6196 062 2.

MCT 35 M.H. VAN EMDEN, *An analysis of complexity*, 1971. ISBN 90 6196 063 0.

MCT 36 J. GRASMAN, *On the birth of boundary layers*, 1971. ISBN 90 6196 064 9.

MCT 37 J.W. DE BAKKER, G.A. BLAAUW, A.J.W. DUIJVESTIJN, E.W. DIJKSTRA, P.J. VAN DER HOUWEN, G.A.M. KAMSTEEG-KEMPER, F.E.J. KRUSEMAN ARETZ, W.L. VAN DER POEL, J.P. SCHAAP-KRUSEMAN, M.V. WILKES & G. ZOUTENDIJK, *MC-25 Informatica Symposium*, 1971. ISBN 90 6196 065 7.

MCT 38 W.A. VERLOREN VAN THEMAAT, *Automatic analysis of Dutch comound words*, 1971. ISBN 90 6196 073 8.

MCT 39 H. BAVINCK, *Jacobi series and approximation*, 1972. ISBN 90 6196 074 6.

MCT 40 H.C. TIJMS, *Analysis of (s,S) inventory models*, 1972. ISBN 90 6196 075 4.

MCT 41 A. VERBEEK, *Superextensions of topological spaces*, 1972. ISBN 90 6196 076 2.

MCT 42 W. VERVAAT, *Success epochs in Bernoulli trials (with applications in number theory)*, 1972. ISBN 90 6196 077 0.

MCT 43 F.H. RUYMGAART, *Asymptotic theory of rank tests for independence*, 1973. ISBN 90 6196 081 9.

MCT 44 H. BART, *Meromorphic operator valued functions*, 1973. ISBN 90 6196 082 7.

MCT 45 A.A. BALKEMA, *Monotone transformations and limit laws*, 1973. ISBN 90 6196 083 5.

MCT 46 R.P. VAN DE RIET, *ABC ALGOL, A portable language for formula manipulation systems, part 1: The language*, 1973. ISBN 90 6196 084 3.

MCT 47 R.P. VAN DE RIET, *ABC ALGOL A portable language for formula manipulation systems part 2: The compiler*, 1973. ISBN 90 6196 085 1.

MCT 48 F.E.J. KRUSEMAN ARETZ, P.J.W. TEN HAGEN & H.L. OUDSHOORN, *In ALGOL 60 compiler in ALGOL 60, Text of the MC-compiler for the EL-X8*, 1973. ISBN 90 6196 086 X.

MCT 49 H. KOK, *Connected orderable spaces*, 1974. ISBN 90 6196 088 6.

* MCT 50 A. VAN WIJNGAARDEN, B.J. MAILLOUX, J.E.L. PECK, C.H.A. KOSTER, M. SINTZOFF, C.H. LINDSEY, L.G.L.T. MEERTENS & R.G. FISKER (eds.), *Revised report on the algorithmic language ALGOL 68*. ISBN 90 6196 089 4.

MCT 51 A. HORDIJK, *Dynamic programming and Markov potential theory*, 1974. ISBN 90 6196 095 9.

MCT 52 P.C. BAAYEN (ed.), *Topological structures*, 1974. ISBN 90 6196 096 7.

MCT 53 M.J. FABER, *Metrizability in generalized ordered spaces*, 1974. ISBN 90 6196 097 5.

MCT 54 H.A. LAUWERIER, *Asymptotic analysis, part 1*, 1974. ISBN 90 6196 098 3.

MCT 55 M. HALL JR. & J.H. VAN LINT (eds.), *Combinatorics, part 1: Theory of designs finite geometry and coding theory*, 1974. ISBN 90 6196 099 1.

MCT 56 M. HALL JR. & J.H. VAN LINT (eds.), *Combinatorics, part 2: Graph theory; foundations, partitions and combinatorial geometry*, 1974. ISBN 90 6196 100 9.

MCT 57 M. HALL JR. & J.H. VAN LINT (eds.), *Combinatorics, part 3: Combinatorial group theory*, 1974. ISBN 90 6196 101 7.

MCT 58 W. ALBERS, *Asymptotic expansions and the deficiency concept in statistics*, 1975. ISBN 90 6196 102 5.

MCT 59 J.L. MIJNHEER, *Sample path properties of stable processes*, 1975. ISBN 90 6196 107 6.

MCT 60 F. GÖBEL, *Queueing models involving buffers*. ISBN 90 6196 108 4.

* MCT 61 P. VAN EMDE BOAS, *Abstract resource-bound classes, part 1*. ISBN 90 6196 109 2.

* MCT 62 P. VAN EMDE BOAS, *Abstract resource-bound classes, part 2*. ISBN 90 6196 110 6.

MCT 63 J.W. DE BAKKER (ed.), *Foundations of computer science*, 1975. ISBN 90 6196 111 4.

MCT 64 W.J. DE SCHIPPER, *Symmetrics closed categories*, 1975. ISBN 90 6196 112 2.

MCT 65 J. DE VRIES, *Topological transformation groups 1 A categorical approach*, 1975. ISBN 90 6196 113 0.

* MCT 66 H.G.J. PIJLS, *Locally convex algebras in spectral theory and eigenfunction expansions*. ISBN 90 6196 114 9.

* MCT 67 H.A. LAUWERIER, *Asymptotic analysis, part 2.* ISBN 90 6196 119 X.

* MCT 68

* MCT 69 J.K. LENSTRA, *Sequencing by enumerative methods.* ISBN 90 6196 125 4.

* MCT 70 W.P. DE ROEVER JR., *Recursive program schemes: semantics and proof theory.* ISBN 90 6196 127 0.

* MCT 71 J.A.E.E. VAN NUNEN, *Contracting Markov decision processes.* ISBN 90 6196 129 7.

* MCT 72 J.K.M. JANSEN, *Simple periodic and nonperiodic Lamé functions and their applications in the theory of eletromagnetism.* ISBN 90 6196 130 0.

An asterisk before the number means "to appear".